BIOCHEMISTRY REVIEW

FIFTH EDITION

Edited by

ERWIN H. MOSBACH, Ph.D.
Chief, Lipid Biochemistry Section
The Public Health Research Institute
of the City of New York, Inc.
New York, N.Y.
Formerly
Associate Professor of Biochemistry
College of Physicians and Surgeons
Columbia University
New York, N.Y.

1800 MULTIPLE CHOICE QUESTIONS AND ANSWERS
—
COMPLETELY REFERENCED

MEDICAL EXAMINATION PUBLISHING COMPANY, INC.
65-36 Fresh Meadow Lane
Flushing, N.Y. 11365

Library of Congress
Catalog Card Number
62-123

ISBN 0-87488-202-8

May 1972

PRINTED IN THE UNITED STATES OF AMERICA

TABLE OF CONTENTS

PREFACE

There are no short cuts to the process of learning! The material on the following pages is not designed to provide the reader with an opportunity to escape from learning basic information that can only be obtained from textbooks. The purpose of this text is to encourage the reader to detect areas of weakness in his understanding of subject matter so that he may return to his texts for a more comprehensive review of the subject. Crossword puzzles are not designed to teach basic English; examination review books do not teach a basic understanding of the subject. However, the following pages will provide an interesting challenge to the student as well as an opportunity to improve his skills with multiple-choice examinations.

The completely referenced form of the question material is provided to facilitate the reader in checking his areas of weakness. Do not compromise by simply looking up the correct answers — the mature approach is to return to your textbook. The author has provided you with a time saving device in rapidly finding a source of information. While testing your memory be aware of the textbook as being the ultimate source of your fund of knowledge. References cited in the individual questions are listed on page 204.

The variations in test questions are offered to familiarize the reader with various types of National Board, State Board and School examinations in use today. The following test material will give you an opportunity to determine your ability to read, digest and comprehend the vast accumulation of available knowledge. Don't be discouraged if you are unable to achieve a perfect test score; indeed be stimulated to learn the material by referring to your textbooks. Successfully solving the problems in this text will surely give you satisfaction.

EACH OF THE QUESTIONS OR INCOMPLETE STATEMENTS BELOW IS FOLLOWED BY FIVE SUGGESTED ANSWERS OR COMPLETIONS. SELECT THE ONE WHICH IS BEST IN EACH CASE:

1. IN NON-COMPETITIVE INHIBITION:
   A. The concentration of active enzyme molecules is reduced
   B. V_{max} is increased above normal
   C. The concentration of active enzyme molecules remains unchanged
   D. K_m is increased
   E. K_m is decreased

   Ref. 2 - p. 128

2. WHICH OF THE FOLLOWING STATEMENTS REGARDING THE ENZYME ASPARTYL CARBOXYLASE IS INCORRECT ?:
   A. The intact enzyme consists of 4 enzyme units and 4 regulatory units
   B. The enzyme catalyzes the reaction of carbamyl phosphate and aspartic acid to form carbamyl aspartase
   C. The enzyme is not inhibited by excess CTP because the latter is bound to the regulatory protein
   D. The enzyme subunit, when separated from the complex, is still active
   E. Aspartyl carbamylase is an allosteric enzyme

   Ref. 1 - p. 244
   Ref. 2 pp. 129-130, 407
   Ref. 3 - pp. 836-839

3. THE NERVE GAS DFP HAS BEEN VERY USEFUL IN THE DETERMINATION OF AMINO ACID SEQUENCES AT THE ACTIVE SITE OF ENZYMES. IT FUNCTIONS AS A NON-COMPETITIVE INHIBITOR OF CERTAIN ESTERASES BY BINDING WITH THE AMINO ACID:
   A. Histidine
   B. Tryptophan
   C. Aspartic acid
   D. Lysine
   E. Serine

   Ref. 2 - pp. 109-110

4. IN COMPETITIVE ENZYME INHIBITION:
   A. The apparent K_m remains unchanged
   B. The apparent K_m is smaller than normal
   C. V_{max} is smaller than normal
   D. V_{max} remains unchanged from normal
   E. The reaction rate is independent of substrate concentration

   Ref. 2 - pp. 126-127

5. KOSHLAND PROPOSED THE CONCEPT OF THE "INDUCED FIT" OF ENZYME ACTION. THIS CONCEPT IMPLIES THAT:
   A. The template or lock-and-key mechanism of enzyme action adequately explains enzyme catalysis
   B. The active site is flexible: the reactive groups of the enzyme protein are brought into proper alignment by the substrate
   C. D- and L- isomers of an optically active substrate react at identical rates
   D. The geometry of the substrate is altered to fit into the enzyme protein
   E. Enzyme-substrate complexes are formed only in oxidation-reduction reactions

   Ref. 2 - pp. 115-116
   Ref. 3 - p. 341

6. IN ENZYME ASSAYS IT IS PREFERABLE TO USE INITIAL VELOCITIES. THIS IS DONE IN ORDER TO:
   A. Make the assay more sensitive
   B. Prevent substrate inhibition
   C. Promote substrate inhibition
   D. Prevent the reverse reaction
   E. Maintain second order kinetics — Ref. 2 - p. 123

7. THE EXPERIMENTALLY DETERMINED VALUE OF K_m:
   A. Depends upon the substrate concentration
   B. Is the same for all substrates of a given enzyme
   C. Depends upon the pH of the assay medium
   D. Has the dimensions of moles/second
   E. Depends upon the enzyme concentration
   Ref. 2 - pp. 125-126

8. IN A SIMPLE ENZYME ASSAY THE SUBSTRATE CONCENTRATION EMPLOYED WAS MUCH LOWER THAN K_m. AS A CONSEQUENCE THE REACTION RATE, ν:
   A. Reached V_{max}
   B. Exhibited zero-order kinetics
   C. Was proportional to substrate concentration
   D. Exhibited no hyperbolic shape even when substrate concentration was increased
   E. Was independent of temperature — Ref. 2 - pp. 121-122

9. WHICH OF THE FOLLOWING STATEMENTS CONCERNING K_m (THE MICHAELIS CONSTANT) IS CORRECT?:
   A. It is equal to the dissociation constant of the enzyme-substrate complex
   B. It is 1/2 of the substrate concentration required to achieve V_{max}
   C. It is identical for all isozymes of an enzyme
   D. It is the substrate concentration required to achieve 1/2 V_{max}
   E. It is independent of the nature of the substrate with which the enzyme reacts — Ref. 2 - p. 125

10. IN BIOTIN-CONTAINING ENZYMES, THE BIOTIN IS VERY TIGHTLY BOUND TO THE ENZYME VIA:
   A. A covalent bond with CO_2
   B. An amide linkage to the epsilon-amino group of lysine
   C. Hydrogen bonding
   D. An amide linkage to the alpha-amino group of lysine
   E. An amide linkage to glutamine — Ref. 3 - p. 423

11. WHICH OF THE FOLLOWING IS AN ESSENTIAL CO-FACTOR IN CARBOXYLATION REACTIONS?:
   A. Coenzyme A
   B. CTP
   C. Lipoic acid
   D. Thiamin pyrophosphate
   E. Biotin
   Ref. 3 - p. 521

12. OXALOACETIC ACID CAN BE FORMED IN ANIMAL TISSUES BY THE CARBOXYLATION OF PYRUVIC ACID. THE ALLOSTERIC EFFECTOR REQUIRED FOR THIS REACTION IS:
   A. Acetyl-CoA
   B. ATP
   C. Biotin
   D. Mg^{++}
   E. GTP
   Ref. 2 - p. 205

13. PYRUVIC DECARBOXYLASE REQUIRES A COENZYME. THIS IS CALLED:
A. Coenzyme A
B. NADP
C. FMN
D. Thiamine pyrophosphate
E. Biotin
Ref. 2 - p. 206

14. THE RATE OF MOST ENZYME-CATALYZED REACTIONS CHANGES WITH pH. AS THE pH INCREASES, THIS RATE:
A. Reaches a minimum, then increases
B. Reaches a maximum, then decreases
C. Increases
D. Decreases
E. None of the above are correct
Ref. 4 - pp. 43-44

15. A SUBSTRATE FOR THE ENZYME ALDOLASE IS:
A. Galactose-6-phosphate
B. Isocitric acid
C. Glucose-1-phosphate
D. Fructose-1, 6-diphosphate
E. Acetaldehyde
Ref. 2 - p. 200

16. COENZYME A CONTAINS THE VITAMIN:
A. Riboflavin
B. Pantothenic acid
C. Pyridoxal
D. Thiamin
E. Niacin
Ref. 1 - p. 326
Ref. 2 - pp. 815-816

17. WHICH OF THE FOLLOWING IS NOT A COMPONENT OF COENZYME A?:
A. Adenylic acid
B. Pantothenic acid
C. Beta-mercaptoethylamine
D. Acetic acid
E. Sulfhydryl groups
Ref. 1 - p. 326

18. SOME BACTERIAL AMINO ACID DECARBOXYLASES REQUIRE A CODECARBOXYLASE WHICH IS:
A. Thiamine pyrophosphate
B. Flavin adenine nucleotide
C. Biotin
D. DPN
E. Pyridoxal phosphate
Ref. 1 - p. 542
Ref. 2 - pp. 812-813

19. MALATE DECARBOXYLASE CATALYZES THE TRANSFORMATION OF MALIC ACID TO PYRUVIC ACID. THIS REACTION INVOLVES:
A. Oxidation + decarboxylation
B. Reduction + decarboxylation
C. Decarboxylation
D. Oxidation
E. Reduction
Ref. 5 - pp. 320-321

20. WHICH CLASS OF ENZYMES CATALYZES THE REACTION, $2H_2O_2 \rightarrow 2H_2O + O_2$?:
A. Dehydrogenases
B. Peroxidases
C. Catalases
D. Hydrolases
E. Oxygenases
Ref. 2 - pp. 250-251

21. CALCULATE THE STANDARD FREE ENERGY CHANGE FOR THE OXIDATION OF NADH BY O_2 AT pH 7 and 25° C:

(E^{o^1} for NADH $\rightarrow NAD^+ + H^+ + 2e$ is 0.320 volts; E^{o^1} for $1/2\ O_2 + 2H^+ + 2e \rightarrow H_2O$ is 0.816 volts. The Faraday constant is 23,060 cal/mole/volt)

A. -21,700 cal/mole
B. +21,700 cal/mole
C. Zero
D. -52,400 cal/mole
E. +52,400 cal/mole

Ref. 5 - p. 201

22. DEHYDROGENASES UTILIZE, AS COENZYMES, ALL OF THE FOLLOWING, EXCEPT:

A. DPN
B. TPN
C. FAD
D. CoA
E. FMN

Ref. 1 - pp. 321-323

23. UREA IS PRODUCED PHYSIOLOGICALLY BY THE ACTION OF THE ENZYME:

A. Urease
B. Glutaminase
C. Arginase
D. Aspartase
E. Uricase

Ref. 2 - pp. 359-361

24. AN ENZYME IN SALIVA WHICH HYDROLYZES STARCH IS:

A. Pepsinogen
B. Chymotrypsin
C. Erepsin
D. Alpha-Amylase
E. Maltase

Ref. 2 - pp. 182-183

25. IF A COENZYME IS REQUIRED IN AN ENZYME REACTION THE FORMER USUALLY HAS THE FUNCTION OF:

A. Acting as an acceptor for one of the cleavage products
B. Enhancing the specificity of the apoenzyme
C. Increasing the number of receptor sites of the apoenzyme
D. Activating the substrate
E. All of the above

Ref. 1 - p. 214

26. THE MICHAELIS-MENTEN HYPOTHESIS:

A. Postulates the formation of an enzyme-substrate complex
B. Enables us to calculate the isoelectric point of an enzyme
C. Postulates that all enzymes are proteins
D. States that the rate of a chemical reaction may be independent of substrate concentration
E. States that the reaction rate is proportional to substrate concentration

Ref. 1 - pp. 225-229
Ref. 2 - pp. 108, 123-126

27. K_m, THE MICHAELIS CONSTANT:

A. Is a true constant since it does not vary when an enzyme acts upon different substrates
B. Is equal to the concentration of substrate which gives the numerical maximal velocity of a given reaction
C. Has a characteristic value for any given enzyme-substrate system which is independent of enzyme concentration
D. Is a measurement of the equilibrium between active and inactive molecules in a given system
E. Is obtained experimentally by measuring the slope of the straight line obtained when plotting V_{max} versus substrate concentration

Ref. 1 - p. 227

EACH GROUP OF QUESTIONS BELOW CONSISTS OF FIVE LETTERED HEADINGS FOLLOWED BY A LIST OF NUMBERED WORDS OR PHRASES. FOR EACH NUMBERED WORD OR PHRASE SELECT THE ONE LETTERED HEADING THAT IS MOST CLOSELY RELATED TO IT:

A. Aerobic oxidases
B. Dehydrogenases
C. Catalases
D. Peroxidases
E. Hydrases

28. ___ $H_2O_2 + AH_2 \rightarrow 2H_2O + A$

29. ___ $R\text{-}CH_2 - \underset{R}{CHOH} \rightleftarrows R\text{-}CH{=}\underset{R}{CH} + H_2O$

30. ___ $AH_2 + 1/2\ O_2 \rightarrow A + H_2O$

31. ___ $AH_2 + B \rightleftarrows A + BH_2$

32. ___ $H_2O_2 \rightarrow 2H_2O + O_2$

Ref. 1 - pp. 200-221, 381-382
Ref. 2 - pp. 132-133

MATCH THE FOLLOWING SERUM ENZYMES WITH THE DISEASE STATES WHICH THEY HAVE BEEN USED TO DIAGNOSE:

A. Pancreatitis
B. Prostatic carcinoma
C. Myocardial infarction
D. Viral hepatitis
E. Obstructive jaundice

33. ___ SGPT (glutamate-pyruvate transaminase)
34. ___ Alkaline phosphatase
35. ___ Amylase
36. ___ Lactic acid dehydrogenase, LDH-1
37. ___ Acid phosphatase

Ref. 2 - pp. 134-136

MATCH EACH ENZYME WITH THE CORRECT COFACTOR OR COENZYME:

A. Fe
B. Cu
C. Zn
D. FAD
E. NAD^+

38. ___ Glucose oxidase
39. ___ Tryosinase
40. ___ Carbonic anhydrase
41. ___ Phosphoglyceraldehyde dehydrogenase
42. ___ Catalase

Ref. 2 - p. 108

A. Chymotrypsin
B. Enteropeptidase
C. Trypsin
D. Carboxypeptidase
E. Pepsin

43. ___ A zinc-containing enzyme which hydrolyzes peptide bonds adjacent to a free carboxyl group
44. ___ An endopeptidase active against peptide bonds formed from alpha-carboxyl group of aromatic amino acids
45. ___ Acts at peptide links formed between alpha-carboxyl of a dicarboxylic amino acid and the alpha-amino radical of an aromatic amino acid
46. ___ Hydrolyzes compounds containing L-lysine or L-arginine residues in which epsilon-amino or guanido grouping is free
47. ___ Specifically converts trypsinogen to trypsin by hydrolysis which releases a hexapeptide

Ref. 1 - pp. 531-534
Ref. 2 - pp. 338-343

MATCH THE ENZYMES TO THEIR CORRECT GROUP DESIGNATION:

A. Oxidoreductase
B. Transferase
C. Hydrolase
D. Lyase
E. Isomerase

48. ___ Transaminase
49. ___ Mutase
50. ___ Peptidase
51. ___ Hydroxylase
52. ___ Aldolase

Ref. 5 - p. 722

A. Invertase
B. Beta-amylase
C. Chymotrypsin
D. Cathepsin
E. Carbonic anhydrase

53. ___ Polysaccharidase
54. ___ Intracellular proteinase
55. ___ Glycosidase
56. ___ Decarboxylase for carbonic acid
57. ___ Endopeptidase

Ref. 1 - pp. 209, 218-219

A. Tetrahydrofolic acid
B. Iron-protoporphyrin
C. Pyridoxal phosphate
D. Flavin adenine dinucleotide
E. Thiamine pyrophosphate

58. ___ One carbon transfer
59. ___ Hydrogen acceptor in aerobic dehydrogenases
60. ___ Cocarboxylase
61. ___ In catalase
62. ___ In transaminases

Ref. 1 - pp. 214-215
Ref. 2 - pp. 246-248, 821

A. Activation energy
B. Non-competitive inhibition
C. Competitive inhibition
D. Van't Hoff equation
E. Zero-order reaction

63. ___ Implies that reaction rate is independent of substrate concentration
64. ___ Expresses relationship between equilibrium constant and temperature
65. ___ Reversible formation of enzyme-inhibitor complex
66. ___ Irreversible combination of enzyme and inhibitor
67. ___ Enzymes decrease it more than inorganic catalysts

Ref. 1 - pp. 230-232, 236-239
Ref. 4 - pp. 43-49

MATCH THE ENZYMES WITH THEIR CORRECT SUBSTRATES:

A. Lipid
B. Carbohydrate
C. Protein
D. RNA
E. DNA

68. ___ Cathepsin
69. ___ Glucan maltohydrolase
70. ___ Lecithinase
71. ___ Ribonuclease
72. ___ Endonuclease

Ref. 1 - pp. 191, 217-221

MATCH THE FOLLOWING ANTIMETABOLITES TO THE CORRESPONDING METABOLITES:

A. Malonic acid
B. Pyridine-3-sulfonic acid
C. Sulfanilamide
D. 6-mercaptopurine
E. Aminopterin

73. ___ Folic acid
74. ___ Succinic acid
75. ___ Adenine
76. ___ Nicotinic acid
77. ___ p-Aminobenzoic acid

Ref. 1 - pp. 237, 240-242
Ref. 3 - pp. 300-302

IN THE FOLLOWING GRAPHS PERTAINING TO THE STUDY OF ENZYME REACTIONS INDICATE THE CORRECT ORDINATES AND ABSCISSAS BY PLACING THE APPROPRIATE LETTERS INTO THE PARENTHESES:

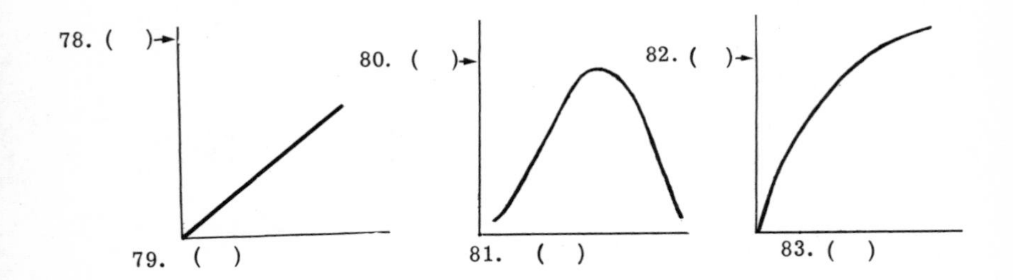

A. pH
B. Enzyme concentration
C. Reaction rate
D. Substrate concentration

Ref. 1 - p. 225
Ref. 2 - pp. 119-122

THE DIAGRAM IS A PLOT OF THE RECIPROCAL OF THE VELOCITY OF AN ENZYMATIC REACTION VERSUS THE RECIPROCAL OF THE SUBSTRATE CONCENTRATION. WHICH CURVE (A, B, OR C) DESCRIBES THE CORRECT SITUATION?:

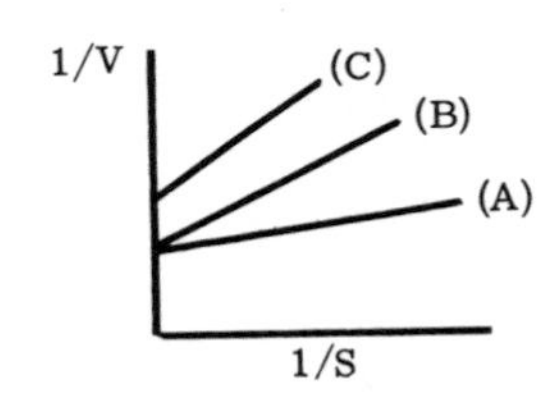

84. ___ Presence of competitive inhibitor
85. ___ No inhibitor
86. ___ Presence of non-competitive inhibitor

Ref. 1 - p. 238
Ref. 2 - pp. 124-128

MATCH THE FOLLOWING ENZYMES WITH THEIR APPROPRIATE COENZYMES:

A. DPN
B. TPN
C. FAD
D. FMN
E. None of these

87. ___ Alcohol dehydrogenase
88. ___ Glyceraldehyde-3-phosphate dehydrogenase
89. ___ D-amino acid oxidase
90. ___ Glucose-6-phosphate dehydrogenase
91. ___ L-amino acid oxidase
92. ___ Muscle aldolase

Ref. 1 - pp. 220, 322-324, 358, 362-365

MATCH THE FOLLOWING ENZYME INHIBITORS WITH THE ENZYME GROUP WHICH COMBINES WITH THE INHIBITOR:

A. Sulfhydryl
B. Serine hydroxyl
C. Fe^{+++}-protoporphyrin enzymes
D. Fe^{+++}, Cu, Zn, other metals
E. Mg, Ca, other metals

93. ___ Cyanide
94. ___ Fluoride
95. ___ Azide
96. ___ N-Ethylmaleimide
97. ___ Diisopropyl phosphofluoridate

Ref. 1 - p. 239
Ref. 3 - pp. 300-302
Ref. 4 - pp. 48-50

THE SUBSTANCES LISTED BELOW HAVE BEEN SHOWN TO ACT AS COMPETITIVE INHIBITORS. MATCH THE INHIBITOR WITH THE ENZYME IT ACTS UPON:

A. Fluorocitrate
B. Oxamate
C. Sulfonamides
D. Pyridine-3-sulfonate
E. Fatty acid amides

98. ___ Alcohol dehydrogenase
99. ___ Biosynthesis of NAD
100. ___ Formation of folic acid
101. ___ L-lactate dehydrogenase
102. ___ Aconitase

Ref. 3 - p. 301

MATCH THE ENZYME INHIBITORS WITH THEIR POSTULATED MODE OF ACTION:

A. Azaserine
B. Pyrithiamine
C. CO
D. Puromycin
E. Avidin

103. ___ Competitive inhibitor of glutamine
104. ___ Competitive inhibitor of thiamine
105. ___ Formation of complex with metallo enzymes
106. ___ Prevents completion of nascent polypeptide chains in protein synthesis
107. ___ Binds to biotin

Ref. 3 - pp. 300-302

MATCH THE SYSTEMATIC NAMES TO THE TRIVIAL NAMES OF THE ENZYMES:

A. Aldolase
B. Alkaline phosphatase
C. Lactate dehydrogenase
D. Cytochrome oxidase
E. Glucokinase

108. ___ L-Lactate: NAD oxidoreductase
109. ___ Orthophosphoric monoester phosphohydrolase
110. ___ ATP-glucose-6-phosphotransferase
111. ___ Ketose-1-phosphate aldehyde-lyase
112. ___ Cytochrome C : oxygen oxidoreductase

Ref. 3 - pp. 327-333

FOR EACH OF THE FOLLOWING NUMBERED ITEMS, CHOOSE THE LETTERED ITEM THAT BEST APPLIES:

A. A specific substance acted upon by an enzyme
B. Dissociation of phosphorylation of ATP from electron transport
C. The oxidative breakdown and release of energy from fuel molecules by reaction with oxygen
D. The energy-yielding enzymatic breakdown of glucose under anaerobic conditions in which lactic acid is the end product
E. The enzymatic phosphorylation of ADP which is coupled to electron transport along the respiratory chain to oxygen

113. ___ Respiration
114. ___ Oxidative phosphorylation
115. ___ Glycolysis
116. ___ Substrate
117. ___ Uncoupling

Ref. 2 - pp. 196, 238, 252, 260-261

A. $2H_2O_2 \rightarrow 2H_2O + O_2$
B. $AH_2 + B \rightarrow A + BH_2$
C. ATP + AMP → 2ADP
D. RH + 1/2 O_2 → ROH
E. L-lactate → D-lactate

118. ___ Kinase
119. ___ Hydroxylase
120. ___ Isomerase
121. ___ Catalase — Ref. 1 - pp. 216-127, 357-360
122. ___ Dehydrogenase — Ref. 3 - pp. 326-333

A. Acetyl-CoA + orthophosphate → CoA + acetylphosphate
B. Citrate + CoA → acetyl-CoA + H_2O + oxaloacetate
C. ATP + acetate + CoA → AMP + pyrophosphate + acetyl-CoA
D. Lactate + NAD^+ → pyruvate + NADH
E. Acetyl-CoA + H_2O → CoA + acetic acid

123. ___ Hydrolase
124. ___ Transferase
125. ___ Ligase
126. ___ Oxidoreductase
127. ___ Lyase — Ref. 3 - pp. 326-333

ANSWER THE FOLLOWING QUESTIONS BY USING THE KEY OUTLINED BELOW:

A. If both the assertion and reason are true statements and the reason is a correct explanation of the assertion
B. If both the assertion and the reason are true statements and the reason is not a correct explanation of the assertion
C. If the assertion is true but the reason is a false statement
D. If the assertion is false but the reason is true
E. If both the assertion and the reason are false

128. Proteolysis is the reverse of protein biosynthesis BECAUSE this is the only way in which the synthesis of species-specific proteins can occur.
Ref. 1 - pp. 534, 661

129. Enzyme and substrate form an activated complex BECAUSE the substrate finds many types of reactive sites on the enzyme.
Ref. 2 - p. 109
Ref. 4 - pp. 44-48

130. Enzymes resemble inorganic chemical catalysts BECAUSE enzymes possess a high degree of specificity.
Ref. 1 - p. 209
Ref. 2 - pp. 113-114

131. A competitive inhibitor decreases the rate of an enzymatic reaction BECAUSE it denatures the enzyme.
Ref. 1 - pp. 236-237

132. The rate of an enzymatic reaction increases as the substrate concentration is increased BECAUSE the concentration of the enzyme-substrate complex increases.
Ref. 1 - pp. 227-229

133. Enzyme-catalyzed reactions are temperature dependent BECAUSE the rate of enzymatic reactions is a function of pH.
Ref. 1 - pp. 231-232, 234-236

134. The active site of chymotrypsin has been shown not to contain a histidine residue BECAUSE treatment of the enzyme with a certain complex ketone (TPCK) results in a parallel loss of activity and of one histidine residue.
Ref. 1 - pp. 254-255

135. It must be assumed that the active site of an enzyme pre-exists before approach of the substrate BECAUSE it has not been possible to show that the conformation of enzymes changes as a consequence of substrate binding.
Ref. 1 - p. 250

136. Diisopropylphosphofluoridate (DFP) inhibits the enzyme chymotrypsin BECAUSE it combines with serine at the active site of the substrate.
Ref. 4 - p. 50

ANSWER THE FOLLOWING QUESTIONS BY USING THE KEY OUTLINED BELOW:

A. If A is greater than B
B. If B is greater than A
C. If A and B are equal or nearly equal

137. A. Rate of mutarotation of α-D-glucose in aqueous solution
B. Rate of mutarotation of α-D-glucose in acetate buffer
Ref. 3 - pp. 355-356

138. A. Molecular weight of pepsin
B. Molecular weight of pepsinogen
Ref. 2 - pp. 339-340

139. A. pH optimum for trypsin
B. pH optimum for pepsin
Ref. 1 - p. 235

140. A. Increase in rate of first order reaction as substrate concentration increases
B. Increase in rate of zero order reactions as substrate concentration increases
Ref. 1 - p. 224

141. A. Activation energy for an enzymatically catalyzed reaction
B. Activation energy for same reaction in absence of enzyme
Ref. 1 - p. 232

142. A. Action of arginase on L-arginine
B. Action of arginase on D-arginine
Ref. 1 - p. 216

143. A. Binding of coenzyme to enzyme protein
B. Binding of prosthetic group to enzyme protein
Ref. 1 - p. 214
Ref. 3 - p. 377

144. A. Enzymatic activity of apoenzyme alone
B. Enzymatic activity of coenzyme alone
Ref. 1 - pp. 214-216

145. A. Enzymatic activity of holoenzyme
B. Enzymatic activity of apoenzyme Ref. 1 - pp. 214-216

146. A. Cold-stability of allosteric enzymes
B. Cold-stability of non-dissociable enzymes
Ref. 3 - p. 303

147. A. K_m if affinity of substrate for enzyme is low
B. K_m if affinity of substrate for enzyme is high
Ref. 5 - p. 137

148. A. Binding of CTP by catalytic subunit of asparate transcarbamylase
B. Binding of CTP by regulatory subunit of asparate transcarbamylase
Ref. 5 - pp. 138-141

149. A. Apparent K_m of an enzyme in the presence of a competitive inhibitor
B. Apparent K_m of an enzyme in the presence of a non-competitive inhibitor
Ref. 5 - pp. 132-134

150. A. K_m of a given enzyme
B. Apparent K_m of the same enzyme in presence of a non-competitive inhibitor
Ref. 5 - pp. 132-134

151. A. K_m of brain hexokinase for glucose
B. K_m of liver glucokinase for glucose
Ref. 5 - p. 137

152. A. V_{max} of rate-limiting enzyme of a given metabolic sequence
B. V_{max} of other enzymes of metabolic sequence
Ref. 5 - pp. 134-136

ANSWER THE FOLLOWING QUESTIONS BY USING THE KEY OUTLINED BELOW:
A. If 1, 2 and 3 are correct
B. If 1 and 3 are correct
C. If 2 and 4 are correct
D. If only 4 is correct
E. If all four are correct

153. CONCERNING THE MICHAELIS-MENTEN HYPOTHESIS:
1. It postulates the existence of an enzyme-substrate complex
2. It asserts that the rate of product formation is independent of the concentration of enzyme-substrate complex
3. It develops a relationship between substrate concentration, enzyme concentration and reaction rate
4. It proves that K_m, the Michaelis constant, is equal to the reaction rate at the maximal reaction velocity Ref. 1 - pp. 225-227

154. IN ENZYMATIC REACTIONS, THE REACTION RATE:
1. Increases as a straight-line function of the substrate concentration
2. Is always first order
3. Is zero order for slow reactions
4. Is a straight-line function of the enzyme concentration
Ref. 1 - pp. 224-230

155. REGARDING THE MICHAELIS CONSTANT K_m:
1. In general, for the same enzyme, the substrate with the greater K_m value is the most reactive
2. For any enzyme-substrate system K_m has a characteristic value which is independent of enzyme concentration
3. It would be very useful to know the value of K_m; unfortunately it cannot be determined experimentally because it is actually a complex constant
4. If the same enzyme can attack several substrates, K_m allows a comparison of the rate of decay of different substrates

Ref. 1 - pp. 227, 231

156. METABOLIC ANTAGONISTS:
1. Exert competitive inhibition when they combine irreversibly with the enzyme
2. Exert competitive inhibition if they combine reversibly with the enzyme
3. Exert non-competitive inhibition by combining with the prosthetic group of the enzyme
4. Exert non-competitive inhibition if there is no relation between degree of inhibition and substrate concentration

Ref. 1 - pp. 236-240

157. CONCERNING ENZYME-SUBSTRATE INTERACTION, IT MAY BE STATED THAT:
1. Enzyme specificity is independent of the conformation of the enzyme protein
2. The failure of enzymes to react with meso-carbon atoms is used as a proof of a multipoint attachment between substrate and enzyme
3. Lactic dehydrogenase cannot bind NADH unless pyruvate is attached to the enzyme
4. Although chymotrypsin has approximately 28 serine residues, only a single one combines with DFP

Ref. 1 - pp. 251, 254

158. THE MECHANISM OF ENZYME ACTION HAS BEEN EXPLAINED IN TERMS OF THE "ACTIVE SITE." THIS MEANS THAT:
1. There may be covalent bonding between enzyme and substrate
2. Functional groups on the enzyme surface participate directly in the reaction mechanism
3. In some cases enzyme activity is due to the specific folding of a peptide chain
4. When enzymes contain no prosthetic groups and require no cofactors the amino acid residues alone are responsible for enzyme activity

Ref. 1 - pp. 258-273

159. THE INFLUENCE OF pH ON ENZYME ACTIVITY IS NOT FULLY UNDERSTOOD. AT PRESENT IT CAN ONLY BE STATED THAT IT IS PROBABLY DUE TO:
1. Dissociation of enzyme and prosthetic group as a function of pH
2. Inactivation of enzyme protein due to pH changes
3. Requirement for specific ionic groupings on enzyme protein for enzyme activity
4. A requirement for the substrate to be in a definite ionic (or non-ionic) form

Ref. 1 - pp. 234-236

160. CONCERNING ALLOSTERIC EFFECTS:
1. Allosteric effectors are usually analogues of the substrate
2. The allosteric site of an enzyme may be quite distant from its active site
3. The allosteric effector catalyzes the denaturation of the polymeric enzyme
4. Most or all allosteric proteins are polymers

Ref. 3 - pp. 299, 303

161. LIPOXYGENASES:
1. Occur largely in plants
2. Occur largely in mammals
3. Catalyze the formation of hydroperoxides
4. Are now known as hydroxylases

Ref. 3 - p. 634

162. CONCERTED ACID-BASE CATALYSIS APPEARS TO BE A MECHANISM COMMON TO MANY ENZYMATIC REACTIONS:
1. It seems to be associated with paired lysyl residues at the active site of the enzyme
2. It proceeds best in solutions containing high concentrations of either acid or base
3. It proceeds best if proton donor and proton acceptor have widely differing dissociation constants
4. It seems to be associated with paired histidyl residues at the active site of the enzyme

Ref. 5 - pp. 109-111
[but see also Ref. 3 - pp. 360-361]

163. IN CO-VALENT CATALYSIS:
1. The attack of the catalyst upon the substrate is usually of the nucleophilic type
2. A co-valently bound substrate-enzyme complex serves as an intermediate
3. A Schiff base frequently serves as an intermediate
4. Histidine appears to play no part in this type of catalysis

Ref. 3 - pp. 362-364

SECTION I - METHODS, BIOLOGICAL OXIDATIONS, BIOENERGETICS

EACH OF THE QUESTIONS OR INCOMPLETE STATEMENTS BELOW IS FOLLOWED BY FIVE SUGGESTED ANSWERS OR COMPLETIONS. SELECT THE ONE WHICH IS BEST IN EACH CASE:

164. FIND THE pH OF A SOLUTION CONTAINING EQUAL CONCENTRATIONS OF A WEAK ACID, HA, $K_a = 2 \times 10^{-5}$, AND THE SODIUM SALT OF THAT ACID, NaA:
A. 4.7
B. 5.3
C. Insufficient information available
D. 3.8
E. 6.7
Ref. 2 - pp. 885-886

165. FIND THE pH OF AN 0.1M SOLUTION OF AN ACID, HA, $K_a = 1 \times 10^{-5}$:
A. 5
B. 4
C. 3
D. 2
E. 1
Ref. 2 - pp. 885-886

166. THE EQUATION

$$pH - pK_a + \log \frac{[salt]}{[acid]} :$$

A. Is generally known as the Henderson-Hasselbalch equation
B. Is incorrect
C. Is useful for strong acids
D. Cannot be used to calculate the pH of buffers
E. Proves that water cannot ionize
Ref. 1 - pp. 98-99

167. WHICH OF THE FOLLOWING IS AN ADEQUATE MARKER ENZYME FOR THE ELECTRON TRANSPORT SYSTEM?:
A. Glucose-6-phosphate dehydrogenase
B. Transhydrogenase
C. Succinate-cytochrome C reductase
D. Malic enzyme
E. Glucose-6-phosphate
Ref. 3 - p. 684

168. ACID-BASE INDICATORS:
A. Are exceptions to the Henderson-Hasselbalch equation
B. Are strong acids with a colorless cation
C. Are strongly dissociated
D. Are salts of weak acids or bases
E. Are colorless in the unionized form
Ref. 1 - pp. 100-101

169. HUMAN SERUM ALBUMIN HAS MORE THAN 100 TITRATABLE ACIDIC GROUPS AND AN APPROXIMATELY EQUAL NUMBER OF TITRATABLE BASIC GROUPS. THIS PROPERTY ALLOWS US TO INFER THAT:
A. Proteins contain several peptide chains
B. Proteins serve as good buffers
C. Enzymatic catalysis is not pH-dependent
D. Proteins have high redox potentials
E. Human serum albumin is an unusual protein
Ref. 1 - pp. 118-119

170. IN THE ELECTRON TRANSPORT CHAIN THE OXIDATION OF ONE MOLE OF NADH TO NAD^+ PRODUCES X MOLES OF ATP. X EQUALS:
A. 3
B. 4
C. 5
D. 6
E. 7
Ref. 1 - p. 343

171. WHICH CYTOCHROME IN THE ELECTRON TRANSPORT CHAIN IS INVOLVED IN THE TERMINAL PROCESS ($1/2\ O_2 \rightarrow H_2O$)?:
A. Cytochrome a
B. Cytochrome b
C. Cytochrome c_1
D. Cytochrome d
E. Cytochrome oxidase
Ref. 2 - p. 249

172. IN BOTH AEROBIC AND ANAEROBIC CELLS THE ENERGY OF CELLULAR OXIDATION REACTIONS IS CONSERVED IN THE COMPOUND:
A. ADP
B. Cyclic AMP
C. Chlorophyll
D. ATP
E. NADP
Ref. 2 - pp. 181, 197, 237-238

173. IF ADMINISTRATION OF COMPOUND "A" RESULTS IN INCREASED EXCRETION OF COMPOUND "B," THEN:
A. A must have stimulated the production of B
B. A must have been converted to B
C. A must have been incorporated into B
D. All of the above statements (A, B, C) are possible
E. None of the above apply
Ref. 1 - p. 289

174. NEGATIVE NITROGEN BALANCE MEANS THAT:
A. No nitrogen is utilized
B. Nitrogen intake equals excretion
C. Nitrogen intake exceeds excretion
D. Nitrogen intake is less than nitrogen excretion
E. All nitrogen is utilized
Ref. 1 - p. 555

175. THE STANDARD FREE ENERGY OF AN OXIDATION REDUCTION REACTION CAN BE CALCULATED IF WE KNOW THE:
A. Activation energy of the two half reactions
B. Absolute temperature at which the reaction takes place
C. Standard oxidation-reduction potentials of the half reactions
D. pH at one of the electrodes
E. Arrhenius constant
Ref. 1 - p. 232
Ref. 4 - pp. 52-55

176. ENZYMES CATALYZING THE ELECTRON TRANSPORT CHAIN ARE PRESENT MOSTLY IN WHICH SUBCELLULAR STRUCTURES?:
A. Endoplasmic reticulum
B. Ribosomes
C. Outer membrane of mitochondria
D. Microsomes
E. Inner membrane of mitochondria
Ref. 3 - pp. 433-435, 682

177. ARRANGE THE PHOSPHATE ESTERS BELOW IN THE MOST CORRECT SEQUENCE, IN ASCENDING ORDER OF STANDARD FREE ENERGIES OF HYDROLYSIS ($-\Delta F^{o}$):
A. Acetyladenylate, S-adenosylmethionine, ATP, glucose-6-phosphate
B. Glucose-6-phosphate, ATP, S-adenosylmethionine, acetyladenylate
C. Glucose-6-phosphate, S-adenosylmethionine, acetyladenylate, ATP
D. ATP, glucose-6-phosphate, acetyladenylate, S-adenosylmethionine
E. Glucose-6-phosphate, S-adenosylmethionine, ATP, acetyladenylate
Ref. 1 - p. 317

178. WHICH OF THE FOLLOWING CAN SERVE AS AN INHIBITOR OF ELECTRON TRANSPORT?:
A. Cyanide
B. Puromycin
C. Actinomycin D
D. Malonate
E. Transhydrogenase
Ref. 3 - p. 686

MATCH THE ENZYME-CATALYZED REACTIONS LISTED BELOW WITH THEIR MOST PROBABLE MECHANISM:

A. General acid-base catalysis
B. Co-valent catalysis
C. Catalysis by distortion

179. ___ Serine → pyruvic acid — Ref. 3 - pp. 363-364, 394
180. ___ Racemization of proline — Ref. 3 - p. 367
181. ___ Glutamate-aspartate transaminase (SGOT) — Ref. 5 - pp. 116-117
182. ___ α-D-glucose ⇄ β-D-glucose — Ref. 3 - pp. 354-360
183. ___ Formation of trigonelline — Ref. 3 - p. 366

MATCH THE CORRECT MARKER ENZYME WITH THE SUBCELLULAR FRACTIONS LISTED BELOW:

A. DNA-dependent RNA polymerase
B. NAD-linked isocitric dehydrogenase
C. β-glucuronidase
D. Glucose-6-phosphatase
E. Glucose-6-phosphate dehydrogenase

184. ___ Mitochondria
185. ___ Cytoplasm
186. ___ Microsomes
187. ___ Nuclei
188. ___ Lysosomes — Ref. 3 - pp. 451-455

SELECT THE ONE BEST ANSWER IN EACH CASE:

189. WHICH OF THE FOLLOWING IS NOT AN INTERMEDIATE IN THE CITRIC ACID CYCLE?:
A. Maleic acid
B. Oxalosuccinic acid
C. Oxaloacetic acid
D. Cis-aconitic acid
E. L-malic acid
Ref. 1 - p. 325
Ref. 2 - pp. 263-264

190. THE LIGHT ENERGY DISSIPATED BY BIOLUMINESCENT ORGANISMS SUCH AS THE FIREFLY IS DERIVED FROM:
A. Coenzyme A
B. ATP
C. NAD
D. NADP
E. Thiamine pyrophosphate
Ref. 1 - p. 291

191. THE FREE ENERGY STORED IN THE ATP MOLECULE CAN BE UTILIZED FOR:
A. Chemical syntheses
B. Heat, osmotic work and mechanical work
C. Electrical work
D. Driving endergonic reactions
E. All of the above
Ref. 1 - pp. 302-303

192. ENTROPY IS A MEASURE OF:
A. The rate of an enzymatic reaction
B. The free energy of an enzymatic reaction
C. The extent to which the total energy of a system is unavailable for the performance of useful work
D. Whether a reaction is exothermic or endothermic
E. The extent of reversibility of a reaction
Ref. 1 - pp. 210, 291, 303

193. THE OXIDATIVE DECARBOXYLATION OF ALL ALPHA-KETO ACIDS IS SIMILAR TO THAT OF PYRUVIC ACID. A COMMON REQUIREMENT FOR ALL IS:
A. NADP
B. Acetyl-CoA
C. Lipoic acid
D. Alpha-ketoglutarate
E. GDP
Ref. 2 - p. 265
Ref. 3 - pp. 390-391

194. IN ANIMAL TISSUE ACETYL-CoA DOES NOT YIELD ACETYL-PHOSPHATE BECAUSE OF THE LACK OF:
A. Condensing enzyme
B. Lipothiamide diphosphate
C. NAD
D. Phosphotransacetylase
E. "Active" acetate
Ref. 1 - p. 414
Ref. 3 - p. 518

195. AN INTERMEDIATE IN "ACTIVE ACETATE" FORMATION FROM FREE ACETATE IS:
A. Acetyl-CoA
B. Acetyl AMP
C. Acetylphosphate
D. Alpha-lipoic acid
E. Lipothiamide diphosphate
Ref. 1 - p. 327
Ref. 4 - p. 87

SECTION I - METHODS, BIOLOGICAL OXIDATIONS, BIOENERGETICS

196. IN THE TRICARBOXYLIC ACID CYCLE CITRIC ACID IS CONVERTED TO ALPHA-KETOGLUTARIC ACID. ALPHA-KETOGLUTARIC ACID HAS A KETONIC OXYGEN IN THE ALPHA POSITION WHEREAS IN CITRIC ACID THE OXYGEN IS IN THE BETA POSITION. INTERMEDIARY STEPS ARE REQUIRED FOR THIS TRANSFORMATION, WHICH IS CATALYZED BY THE ENZYMES:
   A. Isocitric dehydrogenase + oxalosuccinic decarboxylase
   B. Succinic dehydrogenase + aconitase
   C. Aconitase + isocitric dehydrogenase + oxalosuccinic decarboxylase
   D. Oxalosuccinic decarboxylase + succinic dehydrogenase
   E. Condensing enzyme + malic enzyme   Ref. 2 - p. 264

197. WHICH OF THE FOLLOWING IS REQUIRED FOR COENZYME A SYNTHESIS IN LIVER?:
   A. ATP
   B. Cysteine
   C. Pantothenic acid
   D. Mg^{++}
   E. All of these
   Ref. 1 - pp. 642-643

198. THE AEROBIC OXIDATION OF CARBOHYDRATES IN MUSCLE IS ACCELERATED BY CATALYTIC AMOUNTS OF SUCCINIC, FUMARIC, MALIC, OXALOACETIC, ALPHA-KETOGLUTARIC, AND CITRIC ACIDS. THIS EFFECT IS BLOCKED BY MALONATE. THE EFFECT OF MALONATE IS DUE TO:
   A. Inhibition of condensing enzyme
   B. Blocking electron transport by cytochromes
   C. Inhibition of glycolysis
   D. Inhibition of succinic dehydrogenase
   E. Fatty acid synthesis from malonyl-CoA
   Ref. 1 - p. 332
   Ref. 2 - p. 266

199. THE ENZYME ACONITASE CATALYSES THE REVERSIBLE INTERCONVERSION OF CITRIC ACID, CIS-ACONITIC ACID AND ISOCITRIC ACID. AT EQUILIBRIUM CITRIC ACID ACCOUNTS FOR 90% OF THE THREE ACIDS. THE KREBS CYCLE NEVERTHELESS PROCEEDS IN RESPIRING TISSUES BECAUSE:
   A. The citric acid is removed rapidly by condensing enzyme
   B. The 3-point attachment of citric acid removes it as soon as it is produced
   C. The isocitric acid formed is removed rapidly by isocitric acid decarboxylase
   D. The isocitric acid formed is removed rapidly by isocitric acid dehydrogenase
   E. Cis-aconitic acid is removed as soon as formed by cis-aconitic acid dehydrogenase provided Mn^{++} is present
   Ref. 1 - pp. 333-334

200. PYRUVIC ACID CAN BE OXIDIZED VIA THE CITRIC ACID CYCLE. TPP IS REQUIRED TO CHANNEL PYRUVIC ACID INTO THE CYCLE. THE TPP IS USED IN THE SYNTHESIS OF:
   A. Oxalosuccinic acid
   B. Lactic acid
   C. Lipoic acid
   D. Active acetylaldehyde
   E. ATP
   Ref. 2 - pp. 204, 263-266
   Ref. 4 - pp. 87-88

201. IN THE PRESENCE OF CONDENSING ENZYME OXALOACETIC ACID PLUS ACETYL-CoA FORM:
A. Citric acid
B. Alpha-ketoglutaric acid
C. Oxalocitraconic acid
D. Methylglutaconic acid
E. Oxalosuccinic acid
Ref. 2 - p. 263

202. WHICH IS NOT A REACTION OBSERVED IN THE TRICARBOXYLIC ACID CYCLE?:
A. Hydration of cis-aconitate to isocitrate
B. Dehydration of citric acid to form cis-aconitic acid
C. Oxidative decarboxylation of alpha-ketoglutaric acid to form succinyl CoA
D. Hydration of fumaric acid to form malic acid
E. Decarboxylation of citric acid to form oxalosuccinic acid
Ref. 2 - p. 264

203. WHICH STATMENT CONCERNING THE TRICARBOXYLIC ACID CYCLE IS NOT CORRECT?:
A. The oxidation of one molecule of pyruvate requires five atoms of oxygen
B. Carbohydrate, fat and amino acids can be oxidized via the cycle
C. Acetate can only funnel into the cycle as acetyl CoA
D. Only two of the three carbons of pyruvic acid are actually oxidized in one turn of the cycle
E. The cycle occurs only under anaerobic conditions
Ref. 4 - pp. 88-90

204. THE LYSOSOMES:
A. Are commonly known as the Golgi complex
B. Are especially abundant in erythrocytes and brain
C. Are not known to participate in intracellular transport
D. Contain synaptic vesicles
E. Contain cathepsins
Ref. 3 - pp. 437-440

205. A REQUIREMENT FOR "ACTIVE ACETYL" FORMATION FROM PYRUVATE IS:
A. Adenylacetate
B. NADP
C. Lipoic acid
D. Condensing enzyme
E. Phosphotransacetylase
Ref. 3 - p. 519

206. CYTOCHROMES ARE:
A. Riboflavin-containing nucleotides
B. Pyridine nucleotides
C. Iron-porphyrin proteins
D. Metal-containing flavoproteins
E. Peroxidases
Ref. 4 - pp. 82-84

207. RESPIRATORY ACTIVITY, AS EXPRESSED BY Q_{O_2} VALUE, IS HIGHEST FOR WHICH OF THE FOLLOWING?:
A. Mitochondria
B. Cerebral cortex
C. Liver
D. Retina
E. Bone marrow
Ref. 1 - p. 346

208. FLAVOPROTEINS HAVE BEEN SHOWN TO BE INTIMATELY INVOLVED IN ELECTRON TRANSPORT MECHANISM. OF PRIME IMPORTANCE IN THE STRUCTURE OF FLAVOPROTEINS IS:
A. Vitamin A
B. Vitamin B_1
C. Vitamin B_2
D. Nicotinamide
E. Vitamin B_6
Ref. 2 - pp. 245-247, 801

209. AN ESSENTIAL ELEMENT IN THE STRUCTURE OF CYTOCHROME C IS:
A. Mg
B. Co
C. Cu
D. Fe
E. Zn
Ref. 1 - p. 177

210. DEATH DUE TO CYANIDE POISONING RESULTS FROM:
A. Cyanide-hemoglobin complex formation
B. Cyanide combining with the red blood cell
C. Cyanide inhibiting cytochrome oxidase
D. Cyanide inhibiting myoglobin
E. Cyanide blocking oxygen transport in blood
Ref. 1 - p. 751

211. WHICH OF THE FOLLOWING IS LOCATED IN THE MITOCHONDRIA?:
A. Cytochrome oxidase
B. Succinic dehydrogenase
C. Aconitase
D. Phospholipid
E. All of these
Ref. 1 - pp. 336-338

212. THE SELECTIVE PERMEABILITY OF CELL MEMBRANES IS ACHIEVED IN PART BY ACTIVE TRANSPORT SYSTEMS. ACTIVE TRANSPORT DIFFERS FROM PASSIVE TRANSPORT IN THAT THE FORMER:
A. Requires energy but no transport protein
B. Depends primarily on diffusion and osmosis
C. Requires a carrier but no energy
D. Requires energy which is usually provided by high-energy phosphate
E. Is necessarily associated with pinocytosis
Ref. 2 - pp. 14-16

213. THE BIOSYNTHESIS OF COENZYME A REQUIRES DIETARY:
A. Vitamin B_1
B. Mn ions
C. Vitamin B_6
D. Pyruvic kinase
E. Pantothenic acid
Ref. 1 - pp. 642-643

214. A COENZYME REQUIRED IN OXIDATIVE DECARBOXYLATIONS:
A. Coenzyme A
B. Biotin
C. Thiamine diphosphate
D. Coenzyme II
E. Reduced coenzyme I
Ref. 1 - pp. 321-322, 403
Ref. 2 - p. 263

215. THE SPECIFIC SUBSTRATE IN OXIDATIVE PHOSPHORYLATION IS:
A. AMP
B. ADP
C. UDP
D. DPN
E. GDP
Ref. 1 - p. 343
Ref. 2 - pp. 256-258

216. IN THE TRICARBOXYLIC ACID CYCLE ACETATE-1-C^{14} IS INCORPORATED INTO ALPHA-KETOGLUTARATE. THE LATTER IS LABELLED IN ONLY ONE CARBOXYL GROUP. THIS INDICATES THAT THE INTERMEDIATE CITRIC ACID:
A. Is attached to aconitase at 3 points
B. Is asymmetric even though optically inactive
C. Is optically active but not asymmetric
D. Is attached to aconitase at 2 points
E. Cannot be an intermediate because it is symmetrical
Ref. 1 - pp. 333-336

217. UNCOUPLING OF OXIDATIVE PHOSPHORYLATION IMPLIES THAT:
A. The ATPase activity of mitochondria is abolished
B. The mitochondria cease to oxidize citric acid cycle substrates
C. ATP formation ceases but respiration continues
D. ATP formation continues but respiration ceases
E. Cellular activity ceases
Ref. 1 - p. 348

MATCH THE FORMULAS WITH THE CORRECT NAME:

A.

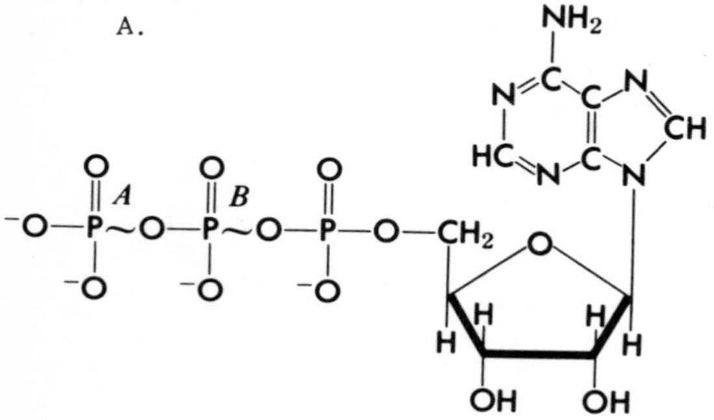

B.

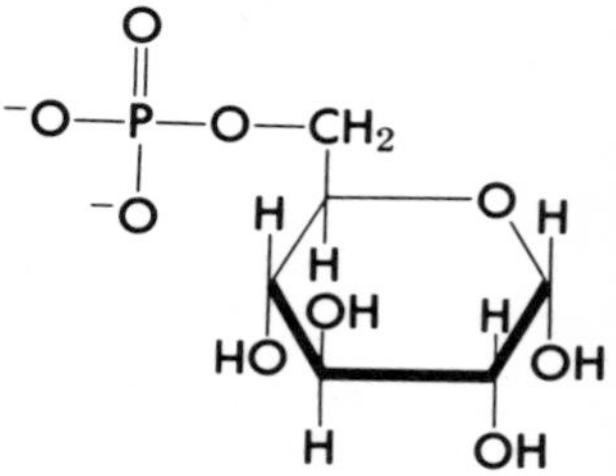

C.

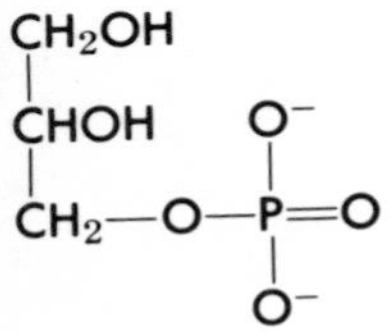

D.

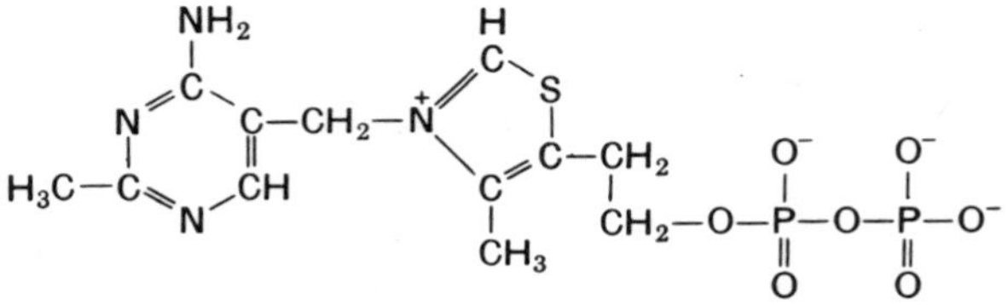

E.

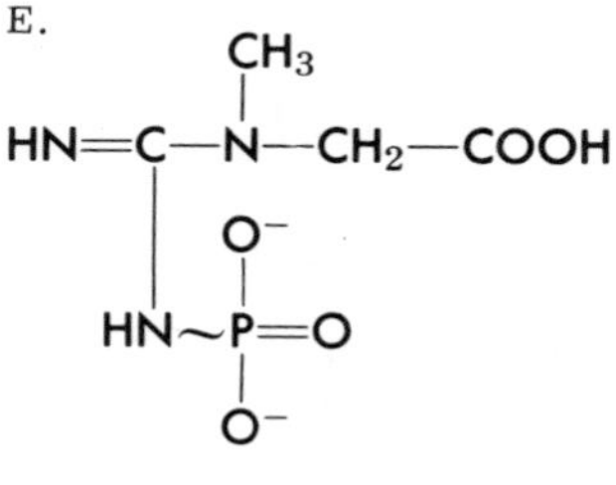

218. ___ Thiamine pyrophosphate

219. ___ ATP

220. ___ α-Glycerophosphate

221. ___ Glucose-6-phosphate

222. ___ Creatine phosphate

Ref. 1 - pp. 314-315; 332

MATCH THE FORMULAS WITH THE CORRECT NAME:

A.

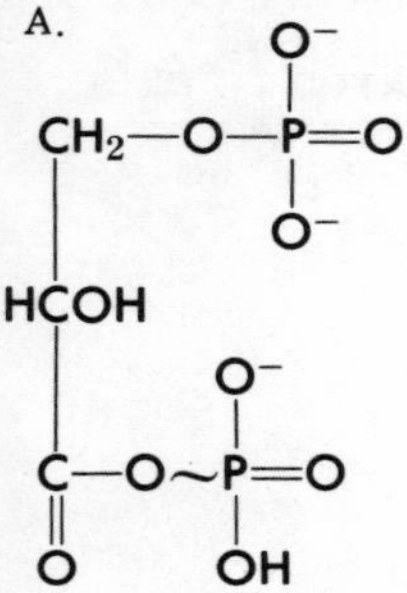

B.

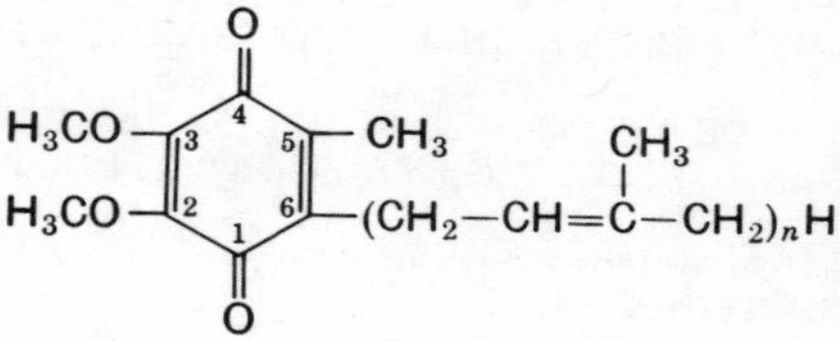

C.

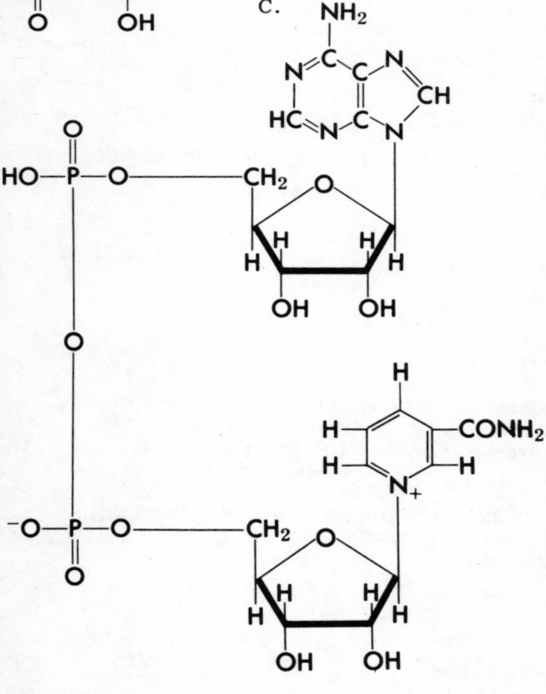

D.

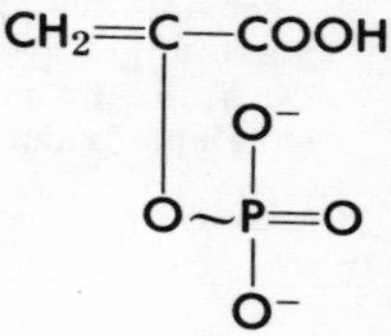

E.

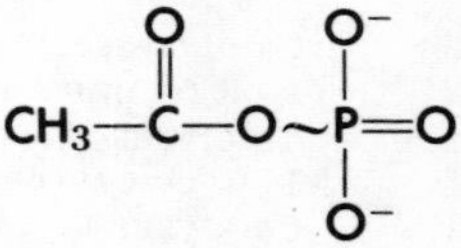

223. ___ Acetyl phosphate

224. ___ NAD

225. ___ Phosphoenolpyruvic acid

226. ___ 1, 3-Diphosphoglyceric acid

227. ___ Ubiquinone

Ref. 1 - pp. 315;322-323

SECTION I - METHODS, BIOLOGICAL OXIDATIONS, BIOENERGETICS

EACH GROUP OF QUESTIONS BELOW CONSIST OF FIVE LETTERED HEADINGS FOLLOWED BY A LIST OF NUMBERED WORDS OR PHRASES. FOR EACH NUMBERED WORK OR PHRASE SELECT THE ONE LETTERED HEADING THAT IS MOST CLOSELY RELATED TO IT:

MATCH THE MONOOXYGENASES LISTED BELOW WITH THEIR CHARACTERISTIC REACTION PRODUCTS:

A. δ-amino-nor-valeraldehyde
B. Tyrosine
C. Corticosterone
D. Lanosterol
E. Norepinephrine

228. ___ Lysine hydroxylase
229. ___ Phenylalanine hydroxylase
230. ___ Steroid-11β-hydroxylase
231. ___ Dopamine hydroxylase
232. ___ Squalene oxidocyclase

Ref. 3 - pp. 638-639

CURRENT CONCEPTS OF THE ELECTRON TRANSPORT CHAIN ASSUME THAT:

A. Ubiquinone
B. Flavoprotein
C. Cytochrome oxidase
D. Cytochrome a
E. Cytochrome b

233. ___ Catalyzes electron transport between flavoprotein and cytochrome b
234. ___ Is at terminal end of the chain
235. ___ Catalyzes electron transport between ubiquinone and cytochrome c
236. ___ Plus cytochrome are components of cytochrome oxidase
237. ___ Catalyzes electron transport between DPNH and ubiquinone

Ref. 1 - pp. 338-341

DIOXYGENASES OCCUR IN BACTERIA AND MAMMALIAN TISSUES. FIND THE CORRECT PRODUCTS WHEN DIOXYGENASES ACT UPON THE FOLLOWING SUBSTRATES:

A. Homogentisic acid
B. L-Tryptophan
C. m-Inositol
D. Unsaturated carboxylic acid
E. Catechol

238. ___ 4-maleylacetoacetic acid
239. ___ Cis, cis-muconic acid
240. ___ L-formylkynurenine
241. ___ Peroxy-acid
242. ___ Glucuronic acid

Ref. 3 - p. 636

MATCH THE FOLLOWING ENZYMES OR CELLULAR COMPONENTS WITH THEIR SUBCELLULAR LOCALIZATION:

A. Lysosomes
B. Nuclei
C. Mistochondria
D. Microsomes
E. Cytosol

243. ___ Electron transport
244. ___ Biosynthesis of NAD
245. ___ Hydrolytic enzymes
246. ___ Biosynthesis of complex lipids
247. ___ Glycolysis Ref. 3 - pp. 450-455

MATCH THE CHARACTERISTIC INHIBITORS OF THE CITRIC ACID CYCLE WITH THE ENZYMES THEY INHIBIT:

A. ATP
B. Fluorocitrate
C. Arsenite
D. β-F-oxalacetate
E. Meso-tartrate

248. ___ Malate dehydrogenase
249. ___ α-oxoglutarate dehydrogenase
250. ___ Fumarase
251. ___ Aconitase
252. ___ Isocitrate dehydrogenase Ref. 3 - p. 608

MATCH THE MONOOXYGENASES LISTED BELOW WITH THEIR CORRECT PROSTHETIC GROUPS:

A. Cu
B. FMN
C. FAD
D. Biopterin
E. P-450

253. ___ L-lysine hydroxylase
254. ___ Lactate decarboxylase (oxidative)
255. ___ Fatty acid desaturase
256. ___ Dopamine hydroxylase
257. ___ Phenylalanine hydroxylase Ref. 3 - pp. 638-639

SECTION I - METHODS, BIOLOGICAL OXIDATIONS, BIOENERGETICS

ANSWER THE FOLLOWING GROUP OF QUESTIONS BY USING THE KEY OUTLINED BELOW:

A. If the item is associated with A only
B. If the item is associated with B only
C. If the item is associated with both A and B
D. If the item is associated with neither A nor B

A. Ubiquinone
B. Cytochrome c
C. Both
D. Neither

258. ___ Participate(s) in electron transport
259. ___ Hemoprotein
260. ___ Flavoprotein
261. ___ Cytochrome oxidase
262. ___ High energy compound

Ref. 1 - pp. 323-324, 340
Ref. 2 - pp. 247-248, 257

A. Xanthine oxidase
B. Diaphorase
C. Both
D. Neither

263. ___ Oxidizes many different aldehydes
264. ___ Catalyzes reduction of methylene blue in presence of DPNH
265. ___ Contains iron, molybdenum and FAD
266. ___ Does not reduce cytochrome c directly
267. ___ Contains TPN as a coenzyme
268. ___ Lipoyl dehydrogenase

Ref. 1 - pp. 220, 366

A. Catalase
B. Peroxidase
C. Both
D. Neither

269. ___ Contains ferriprotoporphyrin IX as a prosthetic group
270. ___ Highest turnover number of any enzyme studied
271. ___ Has four iron atoms per molecule
272. ___ Has one iron atom per molecule
273. ___ Occurs rarely in animal world
274. ___ Hydroperoxidase

Ref. 1 - pp. 179-180, 281-383

A. ATP
B. NAD
C. Both
D. Neither

275. ___ Electron acceptor
276. ___ Contains adenine
277. ___ Contains 2 pyrophosphate bonds
278. ___ Can be further phosphorylated
279. ___ Coenzyme

Ref. 1 - pp. 215, 315, 322, 347

A. Oxidation
B. Reduction
C. Both
D. Neither

280. ___ Succinic acid → fumaric acid
281. ___ Pyruvic acid → lactic acid
282. ___ Hydroquinone → quinone
283. ___ Oxaloacetate → malate
284. ___ Cis-aconitic acid → d-isocitric acid Ref. 1 - pp. 304-307

A. NAD
B. NADP
C. Both
D. Neither

COENZYME OF:
285. ___ Lactate dehydrogenase
286. ___ Glucose-6-phosphate dehydrogenase
287. ___ Alcohol dehydrogenase
288. ___ L-glutamate dehydrogenase
289. ___ Succinate dehydrogenase Ref. 2 - pp. 244-245
Ref. 3 - p. 413

A. NAD
B. NADP
C. Both
D. Neither

290. ___ Coenzymes of dehydrogenases
291. ___ Assay by light absorption of reduced form at 340 nm
292. ___ Assay by light absorption of oxidized form at 340 nm
293. ___ Contains phosphate at C-2' position of ribose
294. ___ Hydride ion transfer Ref. 1 - pp. 322-323, 355-359
Ref. 3 - pp. 154, 410

ANSWER THE FOLLOWING QUESTIONS BY USING THE KEY OUTLINED BELOW:
A. If A is greater than B
B. If B is greater than A
C. If A and B are equal or nearly equal

295. A. Oxidation-reduction potential of cytochrome c
B. Oxidation-reduction potential of flavin nucleotides
Ref. 1 - pp. 307, 340

296. A. Oxidation-reduction potential of pyridine nucleotides
B. Oxidation-reduction potential of flavin nucleotides
Ref. 1 - pp. 307, 340

297. A. Oxidation-reduction potential of water-oxygen system
B. Oxidation-reduction potential of H_2/H^+
Ref. 1 - p. 307

298. A. Liver content of TPN
B. Liver content of DPN Ref. 1 - p. 355

299. A. Free energy of hydrolysis of alpha-glycerophosphate
B. Free energy of hydrolysis of phosphoenolpyruvate
Ref. 1 - p. 317

300. A. Free energy of hydrolysis of glucose-6-phosphate
B. Free energy of hydrolysis of acetyl phosphate
Ref. 1 - p. 317

301. A. ATP yield of the reaction : succinyl CoA → succinate
B. ATP yield of the reaction : alpha-ketoglutarate → succinyl CoA
Ref. 1 - p. 345

302. A. $-\Delta F^{o}$ for hydrolysis of acetyl CoA
B. $-\Delta F^{o}$ for hydrolysis of ATP to ADP and phosphate
Ref. 1 - p. 317

303. A. Diameter of bacterial cells
B. Diameter of mammalian cells Ref. 3 - p. 511

304. A. Dimensions of bacterial cell
B. Dimensions of animal cell mitochondrion
Ref. 3 - p. 511

305. A. Number of nuclei in bacterial cell
B. Number of nuclei in animal hepatic cell
Ref. 3 - p. 511

306. A. Centrifugal force required to sediment microsomes
B. Centrifugal force required to sediment mitochondria
Ref. 3 - p. 448

307. A. Concentration of mixed function oxidases in mitochondria
B. Concentration of mixed function oxidases in microsomes
Ref. 3 - pp. 452-453

308. A. Activity of glucose-6-phosphatase in inner mitochondrial membrane
B. Activity of glucose-6-phosphatase in outer mitochondrial membrane
Ref. 1 - p. 338

309. A. RNA content of mitochondria
B. DNA content of mitochondria Ref. 3 - p. 450

310. A. RNA content of ribosomes
B. RNA content of mitochondria Ref. 3 - pp. 450-452

311. A. Contraction of mitochondria treated with thyroxin
B. Contraction of mitochondria treated with ATP
Ref. 3 - p. 451

312. A. Utilization of one-carbon fragments for anaplerotic reactions
B. Utilization of 2-carbon fragments for anaplerotic reactions
Ref. 3 - p. 492

313. A. Protein synthesis in rough endoplasmic reticulum
B. Protein synthesis in smooth endoplasmic reticulum
Ref. 3 - p. 435

314. A. Percentage of protein in ribosomes
B. Percentage of RNA in ribosomes
Ref. 3 - p. 435

315. A. Relative proportion of DNA in ribosomes
B. Relative proportion of RNA in ribosomes
Ref. 3 - p. 435

316. A. Golgi complexes in secretory cells
B. Golgi complexes in erythrocytes
Ref. 3 - p. 437

317. A. Concentration of hydrolytic enzymes in ribosomes
B. Concentration of hydrolytic enzymes in lysosomes
Ref. 3 - p. 437

318. A. Triglyceride content of mitochondria
B. Phospholipid content of mitochondria
Ref. 3 - pp. 450, 687

319. A. Phosphatidyl choline content of beef heart mitochondria
B. Phosphatidyl ethanolamine content of beef heart mitochondria
Ref. 3 - p. 688

320. A. Activation of mitochondria by highly unsaturated phospholipids
B. Activation of mitochondria by saturated phospholipids
Ref. 3 - p. 688

321. A. Rate of oxidative phosphorylation in mitochondria
B. Rate of oxidative phosphorylation in ribosomes
Ref. 3 - p. 454

322. A. Number of Cu-atoms per mol of ceruloplasmin
B. Number of Cu-atoms per mol of hemocyanin
Ref. 3 - p. 644

323. A. Maximum energy yield when glucose is metabolized by aerobic oxidation
B. Maximum energy yield when glucose is catabolized by anaerobic glycolysis
Ref. 2 - pp. 202-203, 267

SECTION I - METHODS, BIOLOGICAL OXIDATIONS, BIOENERGETICS

ANSWER THE FOLLOWING QUESTIONS BY USING THE KEY OUTLINED BELOW:

A. If 1, 2 and 3 are correct
B. If 1 and 3 are correct
C. If 2 and 4 are correct
D. If only 4 is correct
E. If all four are correct

324. IN THE GLYCOLYTIC PATHWAY OF GLUCOSE CATABOLISM:
1. The breakdown of one molecule of glucose leads to the net formation of 2 molecules of lactic acid and 4 molecules of ATP
2. Since ATP is produced, it is obvious that none is required for intermediate steps
3. Pyruvate phosphokinase is required for the phosphorylation of NAD to NADP
4. Most of the intermediates are phosphoric acid esters which cannot leave the cell

Ref. 2 - pp. 202-203

325.
1. If the reactants in a biological system are maintained in a steady state remote from equilibrium the true value of ΔF for the system may be significantly greater or less than ΔF^o
2. The theory of resonance postulates that, of two possible states, that exhibiting the greater number of resonating forms is the more stable
3. Energy must be supplied if a reaction is to proceed in which the number of resonating forms of a compound is to be minimized
4. Energy-rich phosphates have more resonating forms than low-energy phosphates

Ref. 1 - pp. 316-317

326.
1. The ultimate source of all chemical potential in living systems is the energy of solar radiation
2. Carbon dioxide is formed primarily by direct oxidation of carbon
3. By virtue of the reversibility of the condensing enzyme reaction citric acid can serve as a potential source of active acetyl
4. Fluoroacetate blocks respiration by inhibiting the action of aconitase

Ref. 1 - pp. 302-303; 329-334; 453

327.
1. The highest concentration of cytochromes occurs in heart and hard-working muscles such as breast muscles of birds
2. Tumors contain very little cytochrome
3. Embryonic tissue is very low in cytochrome content
4. Cytochromes are present mainly in the cells of anaerobic organisms

Ref. 1 - pp. 369-372

328. OXIDATIVE PHOSPHORYLATION:
1. Is defined as the formation of ATP in association with oxidative processes
2. Is said to be "uncoupled" when the substrate rather than ADP is phosphorylated
3. Is said to be "uncoupled" when electron transport proceeds at a maximal rate without formation of ATP
4. Requires prior imbibition of water by mitochondria

Ref. 1 - pp. 343-348

329. RESPIRATORY SUBPARTICLES FROM MITOCHONDRIA MAY BE OBTAINED BY THE FOLLOWING PROCEDURES:
1. Treatment with deoxycholate
2. Sonication
3. Treatment with digitonin
4. Treatment with dilute HCl Ref. 3 - p. 681

330. REGARDING METABOLIC PATHWAYS:
1. Malonyl CoA is an intermediate in the biosynthesis of fatty acids, not in their catabolism
2. Anabolic and catabolic pathways are rarely identical
3. The enzymes catalyzing fatty acid anabolism and catabolism are localized in different subcellular compartments
4. It is important that the initial and final reactions of metabolic sequences be easily reversible Ref. 5 - pp. 488-492, 592

331. CONCERNING THE PYRIDINE NUCLEOTIDES:
1. In general, enzymes responsible for oxidations which supply energy to the organism utilize DPN^+, while enzymes which catalyze reductive biosyntheses employ TPN^+
2. When DPN^+ is reduced by 2-electron transfer, the DPN^+ accepts the equivalent of a hydride ion from the oxidized substrate and a proton is liberated to the medium
3. There is no obvious or known relationship between the nature of the substrate and enzyme preference for DPN or TPN
4. The pyridine nucleotides readily dissociate from the dehydrogenases and are really co-substrates which serve as coenzymes only in that there exist <u>other</u> enzymes which catalyze reoxidation of the reduced forms Ref. 1 - pp. 355-359

MATCH THE FORMULAS WITH THE CORRECT NAME:

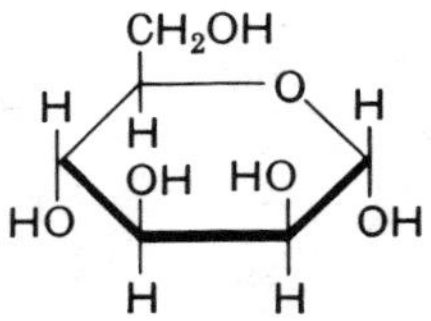

A.

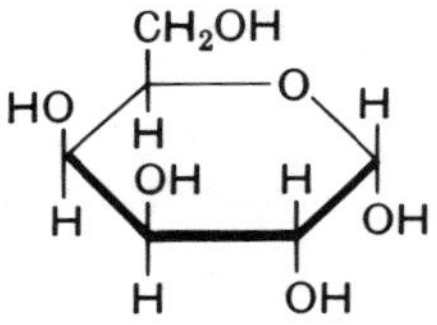

B.

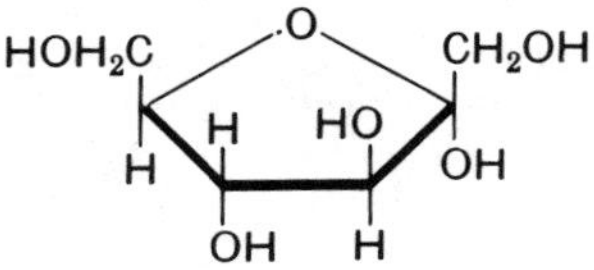

C.

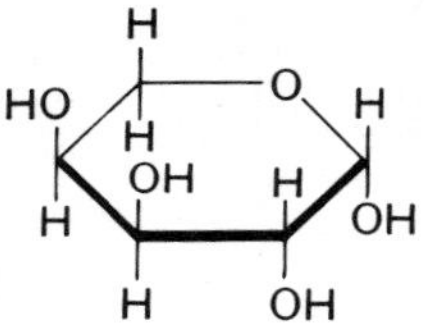

D.

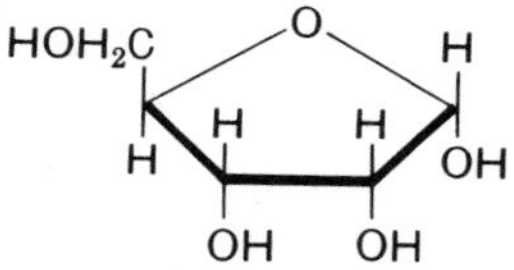

E.

332. ___ α-D-Fructofuranose

333. ___ α-D-Galactopyranose

334. ___ α-D-Ribofuranose

335. ___ α-D-Mannopyranose

336. ___ β-L-Arabinopyranose

Ref. 1 - p. 25

MATCH THE FORMULAS WITH THE CORRECT NAME:

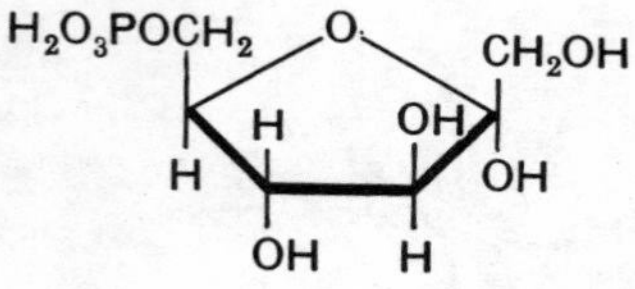

A.

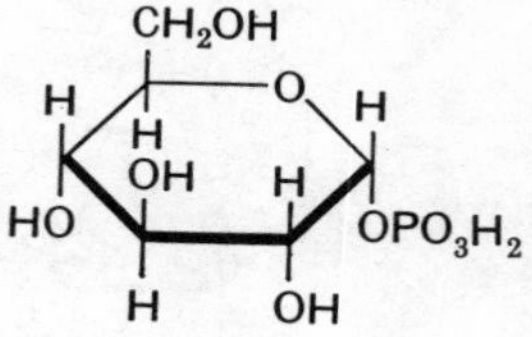

B.

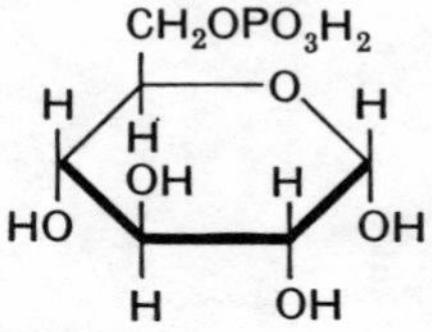

C.

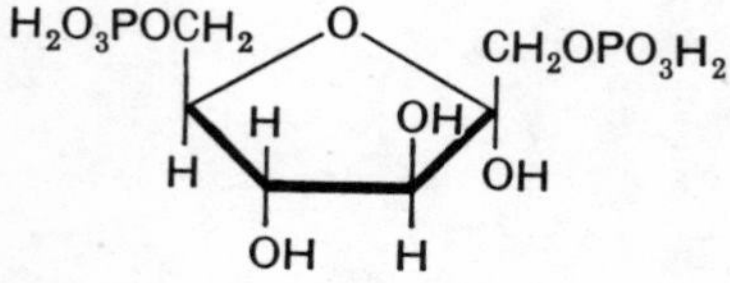

D.

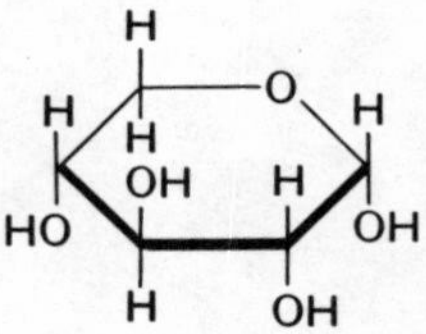

E.

337. ___ Fructose 6-phosphate

338. ___ Glucose 1-phosphate

339. ___ Fructose 1, 6-diphosphate

340. ___ α-D-Xylopyranose

341. ___ Glucose 6-phosphate

Ref. 1 - pp. 26-27

MATCH THE FORMULAS WITH THE CORRECT NAME:

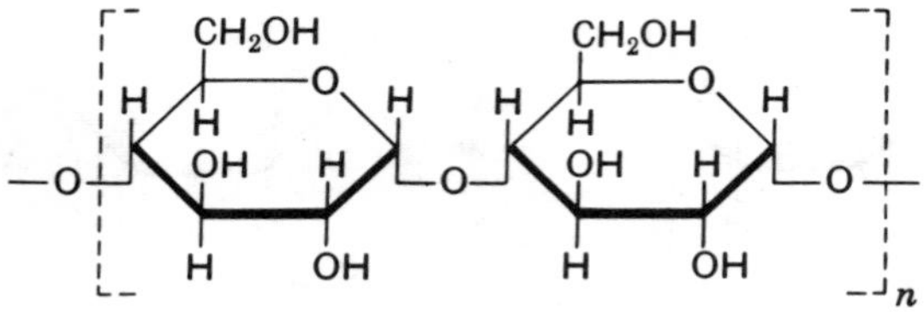

A.

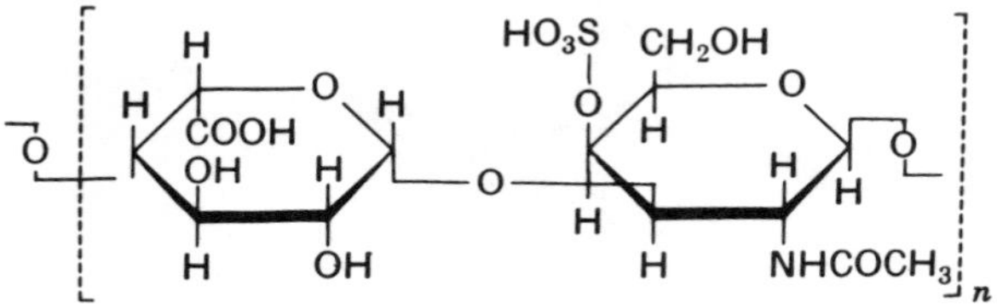

B.

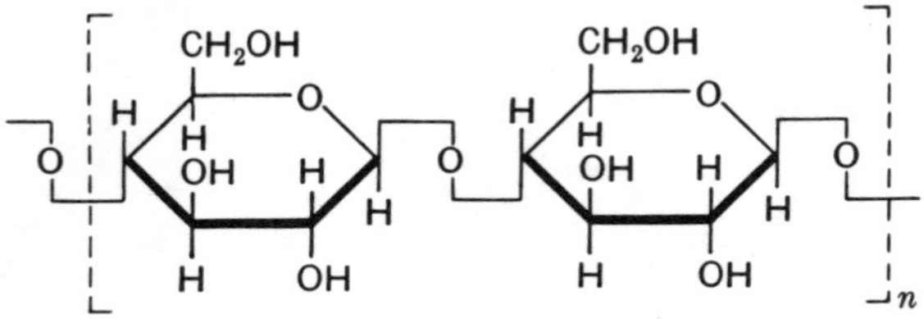

C.

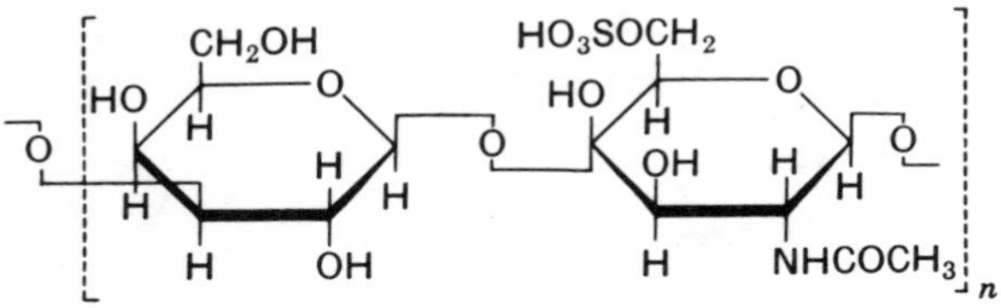

D.

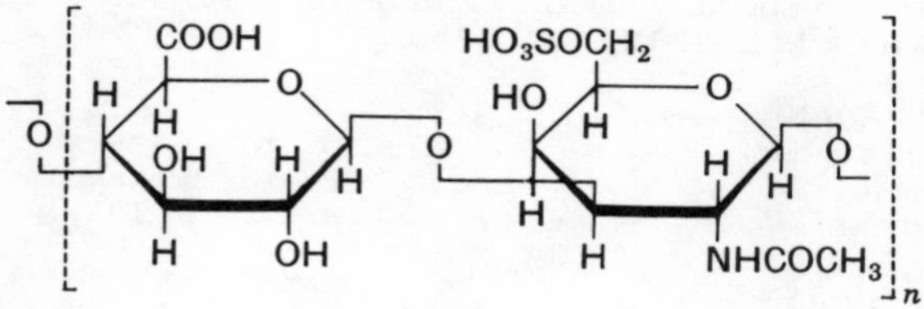

E.

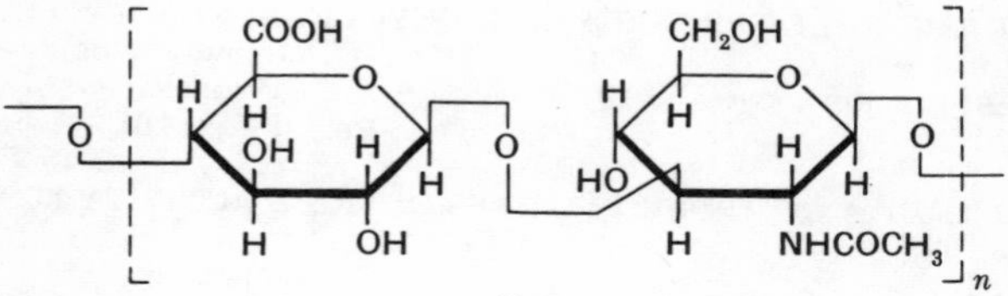

F.

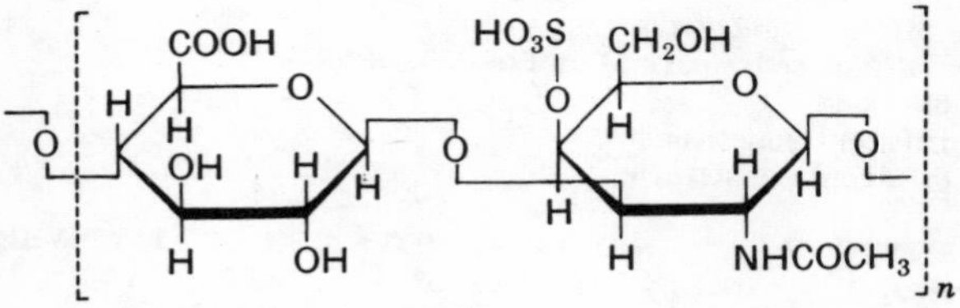

G.

342. ___ Dermatan sulfate

343. ___ Cellulose

344. ___ Chondroitin sulfate A

345. ___ Chondroitin sulfate C

346. ___ Hyaluronic acid

347. ___ Keratosulfate

348. ___ Starch

Ref. 1 - pp. 46; 52-53

CHAPTER II - METABOLISM
SECTION II - CARBOHYDRATES

EACH OF THE QUESTIONS OR INCOMPLETE STATEMENTS BELOW IS FOLLOWED BY 5 SUGGESTED ANSWERS OR COMPLETIONS. SELECT THE ONE WHICH IS BEST IN EACH CASE:

349. GLUCOSE IS A (AN):
A. Oligosaccharide
B. Aldohexose
C. Aldopentose
D. Ketohexose
E. Disaccharide
Ref. 1 - p. 10

350. FRUCTOSE IS A (AN):
A. Oligosaccharide
B. Aldohexose
C. Aldopentose
D. Ketohexose
E. Ketopentose
Ref. 1 - p. 24

351. WHICH OF THE FOLLOWING IS A CARBOCYCLIC ALCOHOL?:
A. Sorbitol
B. β -D-Arabinofuranose
C. Inositol
D. Methyl glucoside
E. Cyclohexane
Ref. 1 - p. 10

352. IN ORDER THAT A COMPOUND POSSESS OPTICAL ACTIVITY IT MUST BE:
A. Colored
B. A carbohydrate
C. Symmetrical
D. Inorganic
E. Asymmetric
Ref. 1 - p. 11

353. A CARBON ATOM IN A MOLECULE BECOMES AN ASYMMETRIC CENTER WHEN IT BEARS:
A. Three different substituents
B. A pair of like and a pair of unlike substituents
C. A double bond
D. Five different substituents
E. Four different substituents
Ref. 1 - p. 12

354. A CERTAIN SUGAR HAD AN OPTICAL ROTATION OF -1.5^{o} WHEN A SOLUTION OF 0.2 g IN 10 ml OF WATER WAS READ IN A 5 cm LONG POLARIMETER TUBE. ITS SPECIFIC ROTATION WAS:
A. -150°
B. $+15^{o}$
C. -1500°
D. -15°
E. $+150^{o}$
Ref. 1 - p. 11

355. A MESO COMPOUND IS OPTICALLY INERT BECAUSE:
A. It is a racemic mixture
B. Resolution methods have not been worked out
C. It is internally compensated
D. It has no atomic centers of asymmetry
E. It cannot be superposed upon its mirror image
Ref. 1 - p. 13

356. THE ARRANGEMENT OF SUGARS INTO D- AND L- CONFIGURATIONS IS BASED UPON THEIR RESEMBLANCE TO D- AND L-:
A. Glyceraldehyde
B. Tartaric acid
C. Glucose
D. Ribose
E. Fructose
Ref. 1 - p. 17

357. GALACTOSEMIA, AN INBORN ERROR OF GALACTOSE METABOLISM, IS A CONSEQUENCE OF THE LACK OF WHICH OF THE FOLLOWING ENZYMES?:
A. UDP-galactose-4-epimerase
B. Uridyl pyrophosphate transferase
C. Galactose-1-phosphatase
D. UDP-galactose transferase
E. Galactokinase
Ref. 2 - pp. 215-216

358. AN EQUILIBRIUM MIXTURE OF GLUCOSE IN WATER CONTAINS α-D-GLUCOPYRANOSE AND β-D-GLUCOPYRANOSE IN A RATIO OF 2:1. THE PREDOMINANCE OF THE β FORM:
A. Is ascribed to the ubiquitous occurrence of mutarotase
B. Is probably due to the fact that the bulkier substituents tend to occupy the axial positions of the molecule
C. Is ascribed to the fact that the bulkier substituents tend to occupy the equatorial positions of the molecule
D. Is due to the tendency of the pyranose ring to open up in polar media such as water
E. Is largely due to a tendency of the molecule to be in the more stable boat form
Ref. 2 - pp. 150-151
Ref. 5 - p. 711

359. IN THE FOLIN-WU QUANTITATIVE METHOD FOR REDUCING SUGARS, A DARK BLUE PRODUCT IS FORMED WHICH CAN BE MEASURED SPECTROPHOTOMETRICALLY. THE COLOR IS DUE TO:
A. Potassium thiocyanate
B. Phosphomolybdous acid
C. Cupric ion
D. Cuprous ion
E. o-Toluidine
Ref. 2 - pp. 152-153
Ref. 5 - p. 711

360. GLUCOSE AND MANNOSE ARE EPIMERS. THIS MEANS THAT:
A. They are mirror images of each other
B. One is an aldose, the other a ketose
C. One is a pyranose, the other a ketose
D. They rotate plane-polarized light in opposite directions
E. They differ only in the configuration of one carbon atom
Ref. 1 - p. 24

361. WHICH OF THE FOLLOWING POLYSACCHARIDES IS NOT A POLYMER OF GLUCOSE?:
A. Amylose
B. Inulin
C. Glycogen
D. Amylopectin
E. Cellulose
Ref. 1 - p. 50

362. HYDROLYSIS OF SUCROSE YIELDS:
A. Glucose only
B. Galactose and glucose
C. Maltose and glucose
D. Fructose and glucose
E. Fructose only
Ref. 1 - p. 41

363. THERE ARE TWO DISTINCT STEREOISOMERIC MODIFICATIONS OF GLUCOSE WHICH ARE INTERCONVERTED IN AQUEOUS SOLUTION TO YIELD AN EQUILIBRIUM MIXTURE. THIS PHENOMENON IS TERMED:
A. Polarization
B. Amphoterism
C. Optical isomerism
D. Mutarotation
E. Conformation
Ref. 1 - p. 19

364. STARCH AND GLYCOGEN BOTH ARE POLYMERS OF:
A. Fructose
B. Glucose-1-phosphate
C. Mannose
D. Glucose
E. Sorbose
Ref. 1 - pp. 46-50

365. THE END PRODUCT OF THE ACID HYDROLYSIS OF GLYCOGEN IS:
A. Amylopectin
B. Dextrin
C. Glucose
D. Maltose
E. Amylose
Ref. 1 - pp. 48-50

366. REDUCTION OF GLUCOSE WITH CALCIUM, IN WATER, PRODUCES:
A. Mannitol
B. Dulcitol
C. Sorbose
D. Glucuronic acid
E. Sorbitol
Ref. 1 - p. 29

MATCH THE FOLLOWING:

A. Conformations
B. Parallel to axis of symmetry of ring
C. Aldehyde + 2 equivalents of an alcohol
D. Ketone + 2 equivalents of an alcohol
E. Extends approximately in plane of ring

367. ___ Axial hydrogen atom
368. ___ Equatorial hydrogen atom
369. ___ Chair and boat forms of cyclohexane
370. ___ Acetal
371. ___ Ketal

Ref. 1 - pp. 22-24
Ref. 2 - pp. 150-152

EACH GROUP OF QUESTIONS BELOW CONSISTS OF FIVE LETTERED HEADINGS FOLLOWED BY A LIST OF NUMBERED WORDS OR PHRASES. FOR EACH NUMBERED WORD OR PHRASE SELECT THE ONE LETTERED HEADING THAT IS MOST CLOSELY RELATED TO IT:

A. Galactosamine
B. Ascorbic acid
C. Glucosamine
D. Glucuronic acid
E. Sialic acid

372. ___ Occurs in human urine bound in glycosidic linkage to various hydroxylated compounds such as estrogens
373. ___ A sugar acid of great biological importance which is widely distributed in nature and which is very readily oxidizable
374. ___ Found in the characteristic polysaccharide of cartilage
375. ___ Occurs in hyaluronic acid and heparin together with glucuronic acid
376. ___ Derivative of neuraminic acid

Ref. 1 - pp. 31-34, 51-55

A. Furfural
B. δ -Gluconolactone
C. Sorbitol
D. Fructose and mannose
E. Sodium gluconate

377. ___ Product of reduction of glucose with H_2 gas under pressure in presence of a metal catalyst
378. ___ Product of dehydration of pentoses by strong mineral acids
379. ___ Product of treatment of glucose with alkaline hypoiodite
380. ___ Product of oxidation of glucose with glucose oxidase
381. ___ Product of treatment of glucose with dilute alkali at low temperature
Ref. 1 - pp. 26-31

IN MAMMALS CERTAIN ENZYMES OF CARBOHYDRATE METABOLISM ARE CONTROLLED BY SPECIFIC REGULATORY SUBSTANCES. MATCH THE ENZYMATIC REACTIONS LISTED BELOW WITH THEIR SPECIFIC ACTIVATORS:

A. Cyclic AMP
B. ADP
C. Fructose-1, 6-diphosphate
D. Acetyl-CoA
E. Glucose-6-phosphate

382. ___ Pyruvate → oxalacetate
383. ___ Fructose-6-phosphate → fructose-1, 6-diphosphate
384. ___ Glucose-1-phosphate → UDPG → glycogen
385. ___ Phosphoenolpyruvate → pyruvate
386. ___ Glycogen → glucose-1-phosphate Ref. 3 - p. 526

A. Feulgen reaction (leucofuchsin)
B. Orcinol
C. Seliwanoff reaction (resorcinol)
D. Naphthoresorcinol
E. Molisch reaction (α-naphthol)

387. ___ Color reaction for deoxypentoses
388. ___ Color test for uronic acids
389. ___ Test for pentoses and uronic acids
390. ___ Test for aldoses and ketoses
391. ___ Color test for ketohexoses Ref. 1 - p. 28

A. 4-O-β-D-galactopyranosyl-D-glucopyranose
B. 4-O-α-D-glucopyranosyl-D-glucopyranose
C. α-D-glucopyranosyl-β-D-fructofuranoside
D. 6-deoxy-L-mannose
E. 6-deoxyl-L-galactose

392. ___ Maltose
393. ___ L-fucose
394. ___ L-rhamnose
395. ___ Lactose
396. ___ Sucrose Ref. 1 - pp. 30, 41-42

ANSWER THE FOLLOWING QUESTIONS BY USING THE KEY OUTLINED BELOW:
A. If A is greater than B
B. If B is greater than A
C. If A and B are equal or nearly equal

397. A. Amount of glucose released by hydrolysis of 1 mole of lactose
B. Amount of glucose released by hydrolysis of 1 mole of maltose
Ref. 1 - pp. 41-42

398. A. Number of α-1, 6-glucosidic bonds in glycogen
B. Number of α-1, 6-glucosidic bonds in cellulose
Ref. 1 - pp. 45, 48-50

399. A. Maltose yield due to action of α-1, 4-glucan maltohydrolase on amylose
B. Maltose yield due to action of α-1, 4-glucan 4-glucanohydrolase on amylose
Ref. 1 - p. 48

400. A. Fructose content of dextrans
B. Glucose content of levans
Ref. 1 - pp. 50, 434

401. A. Number of α-1, 4 linkages in glycogen
B. Number of α-1, 6 linkages in glycogen
Ref. 1 - pp. 48, 437

402. A. Frequency of occurrence of branched chains in starch
B. Frequency of occurrence of branched chains in glycogen
Ref. 1 - pp. 46-50

403. A. Extent of hydrolysis of glycogen by α-1, 4-glucan maltohydrolase
B. Extent of hydrolysis of maltose by α-1, 4-glucan maltohydrolase
Ref. 1 - pp. 46-50

ANSWER THE FOLLOWING QUESTIONS BY USING THE KEY OUTLINED BELOW:
A. If 1, 2 and 3 are correct
B. If 1 and 3 are correct
C. If 2 and 4 are correct
D. If only 4 is correct
E. If all four are correct

404. 1. Glycogen is analyzed for in tissues by solution of the tissue in hot alkali, precipitation of glycogen with alcohol, acid hydrolysis of the precipitate and quantitative analysis of the maltose formed
2. Glycogen is completely hydrolyzed by α-1, 4-glucan maltohydrolase
3. Glycogen is very unstable in hot alkali
4. α-1, 4-Glucan maltohydrolase attacks glycogen, yielding maltose and a limit dextrin
Ref. 1 - pp. 48-50

405. CARBOHYDRATES ARE USUALLY:
1. Colorless liquids
2. Sparingly soluble in organic solvents
3. Non-polar compounds
4. Soluble in water

Ref. 1 - p. 9

406. 1. Yeast maltase attacks only alpha-glucosides
2. Almond emulsin attacks beta-glucosides
3. Invertase catalyzes the hydrolysis of sucrose during which the sign of optical rotation changes from positive to negative
4. Salivary amylase splits β-1, 4-glucosidic bonds

Ref. 1 - pp. 40-41; 47

407. 1. Monosaccharides may be attached to one another by a glycosidic linkage
2. Polysaccharides are high molecular weight polymers of glucose
3. Some polysaccharides when hydrolyzed yield mixtures of hexoses and hexose derivatives
4. Homopolysaccharides are high molecular weight polymers of hexoses and hexose derivatives

Ref. 1 - pp. 43-45

408. 1. Reduction of aldoses or ketoses yields polyhydric alcohols
2. Polysaccharides may contain acetyl, sulfuryl and phosphoryl residues
3. Carbohydrates which posses optical activity have asymmetric molecules
4. Racemic mixtures are optically inactive due to "internal compensation"

Ref. 1 - pp. 11-13; 29; 51-53

409. 1. Lactose is not a reducing sugar
2. Glycosides are formed when a hydroxyl group on the anomeric carbon of a monosaccharide reacts with an alcohol
3. Muramic acid and teichoic acid have been shown to be identical
4. Sucrose is not a reducing sugar

Ref. 1 - pp. 24; 34-35;42;913

MATCH THE FORMULAS WITH THE CORRECT NAME:

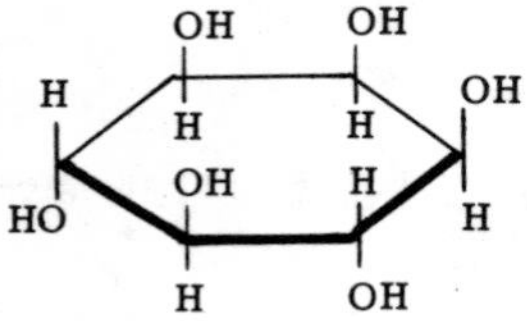

A.

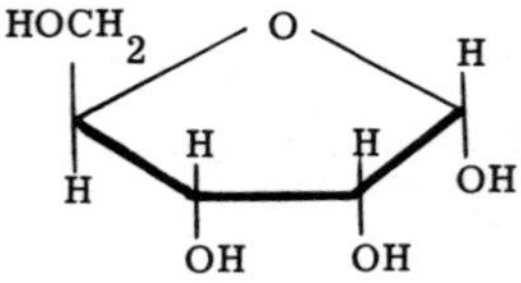

B.

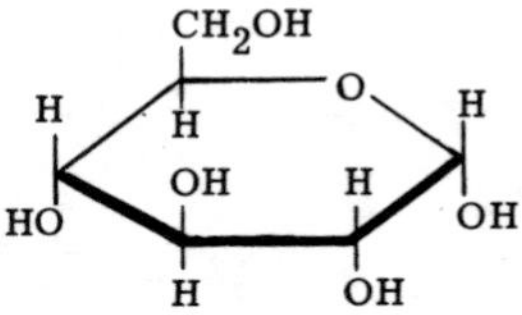

C.

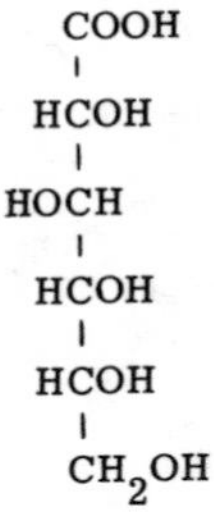

COOH
HCOH
HOCH
HCOH
HCOH
CH_2OH

D.

HC = O
HCOH
HOCH
HCOH
HCOH
COOH

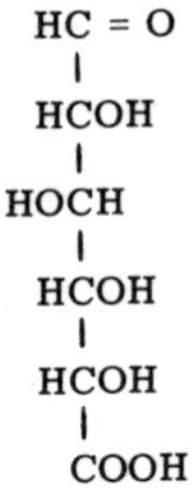

E.

410.___ α-D-ribofuranose

411.___ D-glucuronic acid

412.___ D-gluconic acid

413.___ Myo-inositol

414.___ α-D-glucopyranose

Ref. 1 - pp. 25; 29; 31-32

MATCH THE FORMULAS WITH THE CORRECT NAME

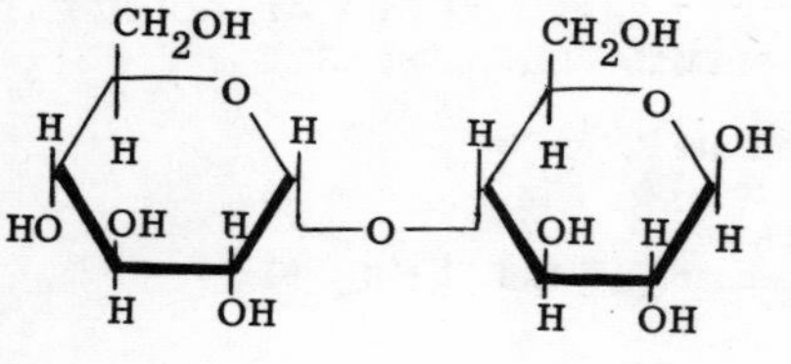

A.

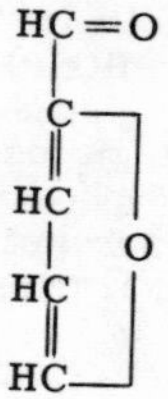

B.

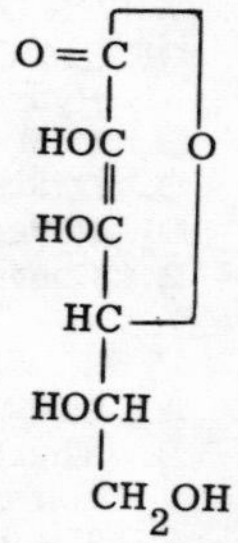

C.

D.

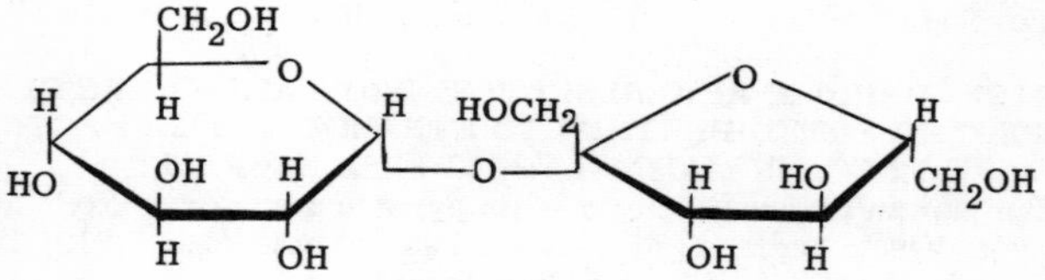

D.

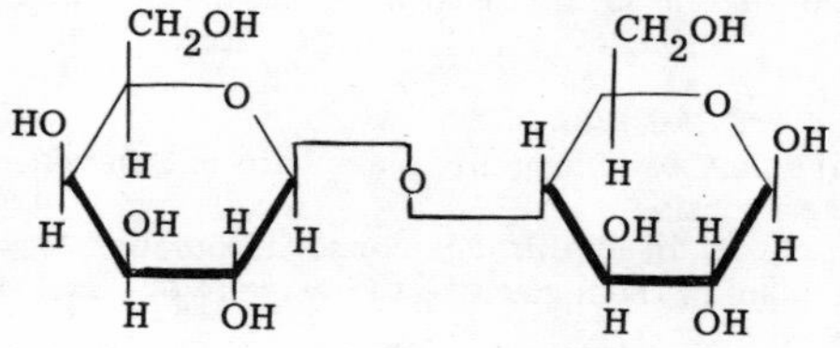

E.

415.___ L-Xyloascorbic acid

416.___ β-Lactose

417.___ β-Maltose

418.___ Sucrose

419.___ Furfural

Ref. 1 - pp. 26, 27; 31, 41-42

SECTION II - CARBOHYDRATES

EACH OF THE QUESTIONS OR INCOMPLETE STATEMENTS BELOW IS FOLLOWED BY 5 SUGGESTED ANSWERS OR COMPLETIONS. SELECT THE ONE WHICH IS BEST IN EACH CASE:

420. WHICH OF THE FOLLOWING IS THE LEAST LIKELY METABOLIC FATE OF GLUCOSE-6-PHOSPHATE IN MAMMALIAN TISSUES?:
A. Hydrolysis to glucose
B. Conversion to glucose-1-phosphate
C. Oxidation to 6-phosphogluconate
D. Reduction to sorbitol-6-phosphate
E. Conversion to fructose-6-phosphate Ref. 1 - p. 391

421. IN ANAEROBIC GLYCOLYSIS:
A. One molecule of glucose is split directly into 2 molecules of lactic acid
B. There is net synthesis of 2 molecules of ATP per molecule of glucose converted to lactate
C. The initial step is the utilization of one ATP molecule in the glucose-6-phosphatase reaction
D. The hexokinase reaction is readily reversible but the phosphofructokinase reaction is not
E. Aldolase catalyzes the reversible isomerization of the hexose phosphates Ref. 1 - p. 392

422. IN YEAST, PYRUVIC ACID RESULTING FROM ANAEROBIC GLYCOLYSIS CAN BE CONVERTED FURTHER TO ETHANOL. THIS REACTION DOES NOT TAKE PLACE IN MAMMALIAN TISSUES BECAUSE:
A. Mammalian tissues do not contain pyruvic acid decarboxylase
B. Acetaldehyde replaces pyruvic acid as the oxidant of DPNH arising from the oxidation of 3-phosphoglyceraldehyde
C. Any acetaldehyde formed is removed as the bisulfite addition product
D. The glycerol which would normally act as precursor is channelled into lipid synthesis
E. The oxygen tension is too great for true anaerobiasis
Ref. 1 - p. 403

423. IN ALCOHOLIC FERMENTATION:
A. The requirement for inorganic phosphate is inherent in the pyruvic acid kinase reaction
B. Arsenite acts as an inhibitor because it removes Mg^{++}
C. The steps leading from glucose to pyruvate are identical to those in animal tissues
D. There is no requirement for P_i; all phosphate is supplied by ATP
E. The formation of ethanol from acetaldehyde is catalyzed by a dehydrogenase which requires thiamine pyrophosphate as a cofactor
Ref. 1 - p. 403

424. THE HEXOSE MONOPHOSPHATE SHUNT IS OF GREAT IMPORTANCE IN CELLULAR METABOLISM BECAUSE IT PRODUCES:
A. NADH
B. ATP
C. CoA
D. ADP
E. NADPH
Ref. 5 - p. 322

425. ONE OF THE SUBSTANCES LISTED BELOW IS NOT AN INTERMEDIATE IN THE FORMATION OF GLUCURONIC ACID FROM GLUCOSE. THE SUBSTANCE IS:
A. L-gulonic acid
B. UDP-glucose
C. Glucose-6-phosphate
D. UDP-glucuronic acid
E. Glucose-1-phosphate
Ref. 1 - p. 423
Ref. 2 - pp. 216-218

426. THE ACTIVITY OF CARBOHYDRATES AS REDUCING AGENTS IS USUALLY DUE TO THE PRESENCE IN THE MOLECULE OF:
A. A hemiacetal
B. A free hydroxyl group
C. At least 2 free hydroxyl groups
D. A free carboxyl group
E. Cupric ions
Ref. 5 - pp. 709-711

427. REVERSAL OF GLYCOLYSIS:
A. Implies operation of the Pasteur effect
B. Implies operation of the Crabtree effect
C. Implies that all steps leading from glucose to lactic acid are readily reversible
D. Requires prior inhibition of the Krebs cycle by malonate
E. Requires participation of enzymes not involved in the normal glycolytic sequence
Ref. 1 - pp. 409-411

428. DEHYDRATION OF 2-PHOSPHOGLYCERIC ACID TO FORM PHOSPHOENOLPYRUVIC ACID:
A. Is catalyzed by phosphoglyceromutase
B. Is inhibited by Mg^{++}
C. Results in an increase in the energy level of the compound
D. Requires ATP to supply the energy for the reaction
E. Is an exergonic reaction
Ref. 1 - p. 401

429. WHICH OF THE FOLLOWING STATEMENTS IS INCORRECT ?:
A. Muscle phosphorylase a contains pyridoxal phosphate
B. Muscle phosphorylase a has twice the molecular weight of muscle phosphorylase b
C. Muscle phosphorylase b is the only active form of the enzyme
D. Muscle phosphorylase b is specifically activated by phosphorylase kinase + ATP
E. The formation of 3', 5'-AMP in muscle is accelerated by epinephrine
Ref. 1 - pp. 440-441
Ref. 2 - pp. 194-195

430. ONE FORM OF GLYCOGEN SYNTHETASE APPEARS TO HAVE A REQUIREMENT FOR GLUCOSE-6-PHOSPHATE. WHICH IS THE BEST EXPLANATION OF THIS PHENOMENON?:
A. It is an example of negative feedback
B. Glucose-6-phosphate is the substrate of the synthetase
C. It determines the specificity of the glycogen branching enzyme
D. It denatures an enzyme which hydrolyzes UDP-glucose
E. It is an example of allosteric activation
Ref. 5 - p. 551

431. PHOSPHOGLYCERALDEHYDE DEHYDROGENASE CONTAINS__________ BOUND TO PROTEIN:
A. TPN^+
B. DPN^+
C. ATP
D. Cu^{++}
E. $AsO_4^{\equiv}$
Ref. 1 - pp. 398-399

432. WHICH OF THE FOLLOWING COMPOUNDS IS FORMED BY PHOSPHOROCLASTIC CLEAVAGE IN E. COLI?:
A. Acetylsulfanilamide
B. Acetyl CoA
C. Citric acid
D. Acetyl phosphate
E. Acetyl choline
Ref. 1 - p. 414

433. CERTAIN MICROORGANISMS SYNTHESIZE SUCROSE BY THE ACTION OF SUCROSE PHOSPHORYLASE UPON:
A. Glucose-1-phosphate + fructose
B. Fructose-6-phosphate + glucose-6-phosphate
C. Glucose + fructose
D. Fructose-6-phosphate + glucose
E. Glucose + fructose + ATP
Ref. 1 - p. 431

434. THE HEREDITARY CHILDHOOD DISEASE GALACTOSEMIA IS CHARACTERIZED BY:
A. High levels of liver UDP-galactose
B. Lack of galactose-1-phosphate
C. Very low galactose-1-P uridyl transferase activity in red cells and liver
D. Unusually rapid reaction of galactose-1-phosphate with uridine diphosphate galactose to form UDP-galactose in liver
E. Lack of UDP-glucose epimerase
Ref. 1 - p. 427

435. UDP-GLUCOSE IS OXIDIZED ENZYMICALLY IN THE PRESENCE OF __________TO YIELD UDP-GLUCURONIC ACID:
A. ATP
B. FMN
C. TPN
D. DPN
E. FAD
Ref. 1 - p. 428

436. WHICH OF THE FOLLOWING BIOCHEMICAL INTERMEDIATES IS CLASSIFIED AS A "HIGH-ENERGY" COMPOUND?:
A. Glucose-6-phosphate
B. Glycerophosphate
C. Glyceraldehyde-3-phosphate
D. 2-phosphoglyceric acid
E. 1,3-diphosphoglyceric acid
Ref. 1 - pp. 314-317

437. AN ENZYME NOT INVOLVED IN GLYCOLYSIS IS:
A. Aldolase
B. Gluconokinase
C. Enolase
D. Pyruvic acid kinase
E. Phosphoglyceromutase
Ref. 1 - p. 390

438. IN PHOTOSYNTHESIS THE REDUCTION OF CO_2 TO CARBOHYDRATE IS:
A. The light reaction
B. An oxygen dependent reaction
C. The dark reaction
D. Independent of the light-catalyzed photolytic cleavage of water
E. Independent of DPNH and ATP
Ref. 1 - pp. 453-460
Ref. 2 - p. 220

439. OXIDATIVE DECARBOXYLATION OF PYRUVATE REQUIRES:
A. Coenzyme A
B. TPN
C. Cytochromes
D. Pyridoxal phosphate
E. Acetyl-CoA
Ref. 1 - pp. 327-328; 403

440. IDIOPATHIC PENTOSURIA IS CHARACTERIZED BY:
A. Excretion of arabinose
B. Excretion of a non-reducing sugar
C. Serious liver damage
D. Excretion of L-xylulose
E. Fructosuria
Ref. 1 - pp. 408; 844

441. WHICH REACTION IS NOT KNOWN TO BE UNDERGONE BY GLUCOSE-6-PHOSPHATE (G-6-P) IN MAMMALIAN TISSUES?:
A. G-6-P ⟶ glucose
B. G-6-P ⟶ G-1-P
C. G-6-P ⟶ 6-phosphogluconic acid
D. G-6-P ⟶ fructose-6-P
E. G-6-P ⟶ 6-phosphoglucuronic acid
Ref. 1 - p. 391

442. N-ACETYLNEURAMINIC ACID IS FOUND IN HUMAN PLASMA, SHEEP SUBMAXILLARY MUCIN AND E. COLI. IT IS SYNTHESIZED FROM N-ACETYLMANNOSAMINE-6-PHOSPHATE AND:
A. Glucuronic acid
B. Uridine diphosphate glucuronic acid
C. Phosphoenolpyruvic acid
D. Uridine triphosphate
E. N-acetylgalactosamine
Ref. 1 - pp. 428-429

443. THE COLOR REACTION OF STARCH WITH AQUEOUS SOLUTION OF IODINE IS LARGELY DUE TO:
A. Amylopectin
B. Achroodextrin
C. Maltose
D. Amylose
E. Free reducing groups of glucose
Ref. 2 - pp. 169-170

444. WHICH OF THE FOLLOWING SUBSTANCES DOES NOT REPRESENT A PATHWAY FOR THE METABOLIC DISPOSAL OF PYRUVATE IN MAMMALS?:
A. Lactic acid
B. Acetyl-CoA
C. Ethanol
D. Alanine
E. Malic acid
Ref. 2 - pp. 204-207

445. THE CONVERSION OF GLUCOSE-6-PHOSPHATE TO FRUCTOSE 1,6-DIPHOSPHATE REQUIRES THE FOLLOWING ENZYMES:
A. Phosphoglucomutase and phosphorylase
B. Phosphoglucomutase and aldolase
C. Phosphohexose isomerase and phosphofructokinase
D. Phosphohexose isomerase and aldolase
E. Glucose-6-phosphatase and a specific pyrophosphorylase
Ref. 1 - pp. 393-394

446. THE HYDROLYSIS OF GLUCOSE-6-PHOSPHATE IS CATALYZED BY A SPECIFIC PHOSPHATASE WHICH IS FOUND ONLY IN:
A. Liver, intestine and kidney
B. Brain, spleen and adrenals
C. Striated muscle
D. Plasma
E. Red cells
Ref. 1 - p. 391

447. THE INHIBITION OF GLYCOLYSIS BY OXYGEN IS KNOWN AS:
A. The Crabtree effect
B. The glycolytic effect
C. The Pasteur effect
D. The fermentation phenomenon
E. The Hill reaction
Ref. 1 - pp. 411-412

448. BEFORE PYRUVIC ACID ENTERS THE TRICARBOXYLIC ACID CYCLE IT MUST BE CONVERTED TO:
A. Acetyl-CoA
B. Lactate
C. Alpha-ketoglutarate
D. Fumarate
E. Citrate
Ref. 1 - pp. 327-328

449. THE FOLLOWING ENZYME IS FOUND IN "THE HEXOSE MONOPHOSPHATE SHUNT":
A. Glucose-6-phosphatase
B. Phosphorylase
C. Hexokinase
D. Glucose-6-phosphate dehydrogenase
E. Malic enzyme
Ref. 1 - pp. 415; 419

450. THE CLEAVAGE OF FRUCTOSE-1,6-DIPHOSPHATE DURING GLYCOLYSIS IS CATALYZED BY THE ENZYME:
A. Fructokinase
B. Diphosphofructose phosphatase
C. Aldolase
D. Phosphofructokinase
E. Enolase
Ref. 2 - p. 198

451. DEHYDROGENASE ENZYMES OF THE HEXOSE MONOPHOSPHATE SHUNT ARE:
A. DPN specific
B. TPN specific
C. FAD specific
D. FMN specific
E. Thiamin pyrophosphate specific
Ref. 1 - p. 419

452. IN THE LIVER GLUCOSE IS CONVERTED IN LARGE PART TO GLYCOGEN. AN ESSENTIAL SUBSTANCE FOR THIS CONVERSION IS:
A. Pyruvate kinase
B. Cytidine triphosphate
C. Glucose-phosphate dehydrogenase
D. Uridine triphosphate
E. Guanosine
Ref. 1 - pp. 425, 434, 436

453 WHICH OF THE FOLLOWING ENZYME ACTIVITIES IS NOT DECREASED IN UNCONTROLLED DIABETES?:
A. Glucokinase
B. Pyruvate carboxylase
C. Acetyl-CoA carboxylase
D. Citrate cleavage enzyme
E. NADP-malate dehydrogenase
Ref. 5 - p. 561

454. WHICH OF THE FOLLOWING ENZYME ACTIVITIES IS NOT INCREASED IN RESPONSE TO HIGH GLUCOSE INTAKES?:
A. Pancreatic amylase
B. Glucose-6-phosphate dehydrogenase
C. Acetyl-CoA carboxylase
D. Hepatic glucose-6-phosphatase
E. Citrate cleavage enzyme
Ref. 5 - p. 550

455. FOR THE COMPLETE OXIDATION OF ONE MOLE OF GLUCOSE VIA THE CITRIC ACID CYCLE, HOW MANY ENERGY-RICH BONDS IN THE FORM OF ATP ARE PRODUCED?:
A. 0
B. 2
C. 12
D. 24
E. 36
Ref. 5 - p. 260

456. TISSUES FORM LACTIC ACID FROM GLUCOSE. THIS PHENOMENON IS TERMED:
A. Aerobic glycolysis
B. Anaplerosis
C. Oxidation
D. Oxidative phosphorylation
E. Anaerobic glycolysis
Ref. 2 - p. 197

457. THE RATE OF ABSORPTION OF SUGARS BY THE SMALL INTESTINE IS GREATEST FOR:
A. Pentoses
B. Hexoses
C. Polysaccharides
D. Disaccharides
E. Oligosaccharides
Ref. 1 - p. 388

458. GLYCOGEN IS CONVERTED TO GLUCOSE-1-PHOSPHATE BY:
A. UDPG-transferase
B. Branching enzyme
C. Phosphorylase
D. Dephosphophosphorylase kinase
E. Isomerase
Ref. 1 - p. 438

459. THE ENZYME WHICH HYDROLYZES SUCROSE TO GLUCOSE AND FRUCTOSE IS:
A. Emulsin
B. Amylase
C. Invertase
D. Saccharase
E. Sucrose phosphorylase
Ref. 1 - p. 41

460. A REQUIREMENT FOR CONVERSION OF FRUCTOSE-6-PHOSPHATE TO FRUCTOSE-1, 6-DIPHOSPHATE:
A. ADP
B. TPN
C. Hexokinase
D. ATP
E. Fructose-1-PO_4
Ref. 1 - pp. 394-396

461. RIBOSE-5-PHOSPHATE + XYLULOSE-5-PHOSPHATE $\rightleftarrows$ SEDOHEPTULOSE-7-PHOSPHATE + GLYCERALDEHYDE-3-PHOSPHATE. THIS IS AN EXAMPLE OF WHICH TYPE OF ENZYMATIC REACTION?:
A. Transphosphorylation
B. Transketolation
C. Transamidination
D. Transamination
E. Transglycosylation
Ref. 1 - pp. 416-417

462. UNDER ANAEROBIC CONDITIONS THE GLYCOLYSIS OF ONE MOLE OF GLUCOSE YIELDS:
A. One mole of ATP
B. Two moles of ATP
C. Eight moles of ATP
D. Thirty moles of ATP
E. No ATP
Ref. 1 - p. 392

463. IN THE REACTION ATP + GLUCOSE $\rightarrow$ ADP + X; THE X IS:
A. Glucose-6-phosphate
B. Glucose-1-phosphate
C. Inorganic phosphate
D. Fructose
E. Fructose-6-phosphate
Ref. 1 - p. 389

464. THE REACTION ATP + GLUCOSE $\rightarrow$ ADP + X IS CATALYZED BY:
A. Aldolase
B. Phosphohexose isomerase
C. Phosphorylase
D. Fructokinase
E. Hexokinase
Ref. 1 - p. 389

465. THE REACTION ATP + GLUCOSE $\rightarrow$ ADP + X HAS A REQUIREMENT FOR:
A. Mn^{++}
B. Primer
C. Insulin
D. Mg^{++}
E. Inorganic phosphate
Ref. 1 - p. 389

EACH GROUP OF QUESTIONS BELOW CONSISTS OF FIVE LETTERED HEADINGS FOLLOWED BY A LIST OF NUMBERED WORDS OR PHRASES. FOR EACH NUMBERED WORD OR PHRASE SELECT THE ONE LETTERED HEADING THAT IS MOST CLOSELY RELATED TO IT:

A. Glucose-6-phosphate dehydrogenase
B. Phosphopentose epimerase
C. Transketolase
D. Transaldolase
E. Phosphopentose isomerase

466. ___ Catalyzes formation of the 7-carbon sugar sedoheptulose
467. ___ Catalyzes interconversion of ketopentoses and aldopentoses
468. ___ Generates TPNH
469. ___ Transfers carbons 1, 2 and 3 of a ketose phosphate to carbon-1 of an aldose phosphate
470. ___ Catalyzes epimerization about carbon-3 of pentose phosphates

Ref. 1 - pp. 415-419

IN MAMMALS CERTAIN ENZYMES OF CARBOHYDRATE METABOLISM ARE CONTROLLED BY SPECIFIC REGULATORY SUBSTANCES. MATCH THE ENZYMATIC REACTIONS LISTED BELOW WITH THEIR SPECIFIC INHIBITORS:

A. Citrate
B. ATP
C. PP_i
D. Glucose-6-phosphate
E. Malonyl-CoA

471. ___ Glucose → glucose-6-phosphate
472. ___ Gructose-6-phosphate → fructose-1, 6-diphosphate
473. ___ Pyruvate → oxalacetate → phosphoenolpyruvate
474. ___ Glucose-6-phosphate → glucose
475. ___ Phophoenolpyruvate → pyruvate

Ref. 3 - p. 526

A. Glycogen synthetase
B. Glycogen branching enzyme
C. Amylo-(1, 6)-glucosidase
D. Phosphorylase kinase
E. Glucose-6-phosphatase

476. ___ Not present in muscle
477. ___ Reactivates inactive liver phosphorylase
478. ___ Forms alpha-1, 4-linkages in glycogen
479. ___ Glycogen debranching enzyme
480. ___ Forms alpha-1, 6-linkages in glycogen

Ref. 1 - pp. 436-438, 441-443
Ref. 2 - pp. 190-195

A. Ferredoxin
B. Chloroplast
C. Ribulose-1, 5-diphosphate
D. Grana
E. Chromatophore

481. ___ Primary CO_2 acceptor in photosynthesis
482. ___ Photosynthetic pyridine nucleotide reductase
483. ___ Contains enzymes required for CO_2 fixation
484. ___ "Photosynthetic unit" of bacteria
485. ___ "Photosynthetic unit" of chloroplasts Ref. 1 - pp. 367-368, 453, 458-459

ANSWER THE FOLLOWING QUESTIONS BY USING THE KEY OUTLINED BELOW:
A. If A is greater than B
B. If B is greater than A
C. If A and B are equal or nearly equal

486. A. Molecular weight of muscle phosphorylase a
B. Molecular weight of muscle phosphorylase b
Ref. 1 - p. 440

487. A. Decarboxylation of pyruvic acid to form acetaldehyde in plants
B. Decarboxylation of pyruvic acid to form acetaldehyde in animal tissues
Ref. 1 - p. 403

488. A. (DPNH)/(DPN) ratio in cells under aerobic conditions
B. (DPNH)/(DPN) ratio in cells under anaerobic conditions
Ref. 1 - p. 402

489. A. Number of alpha-1, 4-linkages in glycogen
B. Number of alpha-1, 6-linkages in glycogen
Ref. 2 - pp. 190-191
Ref. 3 - p. 549

490. A. Lipid storage in plants
B. Carbohydrate storage in plants Ref. 1 - p. 442

491. A. Requirement of TPNH for CO_2 fixation by pyruvic acid carboxylase
B. Requirement of TPNH for CO_2 fixation by malic enzyme
Ref. 1 - p. 408

492. A. Carbohydrate storage in mammals
B. Lipid storage in mammals Ref. 1 - p. 442

493. A. Glycogen breakdown by hydrolysis in liver
B. Glycogen breakdown by phosphorolysis in intestine
Ref. 1 - p. 443

494. A. Blood sugar level before breakfast
B. Blood sugar level after lunch Ref. 2 - p. 188

495. A. Blood sugar level in diabetes mellitus
B. Blood sugar level in renal glucosuria

Ref. 2 - pp. 222-223

ANSWER THE FOLLOWING QUESTIONS BY USING THE KEY OUTLINED BELOW:

A. If an increase in the first is accompanied by an increase in the second or if a decrease in the first is accompanied by a decrease in the second
B. If an increase in the first is accompanied by a decrease in the second or if a decrease in the first is accompanied by an increase in the second
C. If changes in one do not affect the other

496. 1. Blood concentration of epinephrine
2. Glycogen concentration in liver

Ref. 1 - pp. 443-444

497. 1. Blood concentration of 11-oxysteroids
2. Glycogen concentration in liver

Ref. 1 - p. 444

498. 1. Blood thyroxin concentration
2. Liver glycogen concentration

Ref. 1 - p. 444

499. 1. Concentration of blood insulin (but less than convulsive levels)
2. Concentration of muscle glycogen

Ref. 1 - p. 444

500. 1. Glucagon concentration in blood
2. Glycogen concentration in liver

Ref. 1 - p. 444

501. 1. Secretion of glucagon by alpha-cells
2. Conversion of phosphorylase b to phosphorylase a

Ref. 1 - p. 444

502. 1. Blood concentration of epinephrine
2. Blood concentration of lactate

Ref. 1 - p. 449

503. 1. Blood concentration of insulin
2. Glycogen content of heart muscle

Ref. 1 - p. 444

ANSWER THE FOLLOWING QUESTIONS BY USING THE KEY OUTLINED BELOW:

A. If 1, 2 and 3 are correct
B. If 1 and 3 are correct
C. If 2 and 4 are correct
D. If only 4 is correct
E. If all four are correct

504. REACTIVATION OF INACTIVE LIVER PHOSPHORYLASE IS FAVORED BY:
1. Insulin
2. Epinephrine
3. ACTH
4. Glucagon

Ref. 1 - p. 444

505.
1. Phosphoglucomutase becomes phosphorylated during the reaction which it catalyzes: glucose-1-phosphate $\rightleftharpoons$ glucose-6-phosphate
2. The equilibrium of the phosphohexose isomerase reaction favors the formation of glucose-6-phosphate over that of fructose-6-phosphate
3. Brain and yeast contain a hexokinase specific for fructose
4. The conversion of fructose-6-phosphate to fructose-1, 6-diphosphate requires ATP

Ref. 1 - pp. 389;395;424

506.
1. Reduction of dihydroxyacetone phosphate by DPNH results in the synthesis of a compound which is utilized for the synthesis of phosphatidic acid
2. Glycerol can arise from the action of a phosphatase on alpha-glycerophosphate
3. Inorganic phosphate is important in glycolysis by virtue of its requirement for synthesis of 1, 3-diphosphoglyceric acid
4. Phosphoglyceraldehyde dehydrogenase contains bound cysteine and TPN^+ which are required for its activity

Ref. 1 - pp. 398; 403

507. THE PHOSPHOGLUCONATE SHUNT IS IMPORTANT BECAUSE:
1. It can completely oxidize glucose independently of the citric acid cycle
2. It generates TPNH which is necessary for fatty acid synthesis
3. It produces ribose, and 4-carbon and 7-carbon sugars
4. It produces more ATP for combustion of glucose than the glycolytic pathway and citric acid cycle

Ref. 1 - p. 415

508.
1. Glycogen phosphorylase contains pyridoxal phosphate
2. Muscle phosphorylase b has a molecular weight 1/2 that of phosphorylase a
3. Muscle phosphorylase b can be activated by AMP
4. Reactivation of inactive liver phosphorylase involves phosphorylase kinase

Ref. 1 - pp. 440-441

509. IN DIABETES:
1. Glucose-6-phosphatase activity increases
2. Transformation of glucose to intracellular glucose-6-phosphate is inhibited
3. Glycogen synthesis is decreased
4. Rate of glucose utilization is no longer dependent upon blood concentration

Ref. 1 - pp. 447-451

510. THE PHOSPHOGLUCONATE OXIDATIVE PATHWAY CAN BE CONSIDERED A CYCLIC PROCESS, PROVIDED 4 GLYCOLYTIC ENZYMES PARTICIPATE:
1. These enzymes are triosephosphate isomerase, aldolase, diphosphofructose phosphatase and hexose phosphate dehydrogenase
2. During this cyclic process 6 moles of heptosephosphate are converted to 5 moles of hexosephosphate and 6 moles of CO_2
3. The usefulness of the pathway is that it provides a mechanism for producing ribose and NADH
4. In liver approximately 30% of CO_2 arising from glucose is contributed by this pathway

Ref. 1 - p. 418

511. GLYCOGEN STORAGE IN LIVER IS UNDER HORMONAL CONTROL. SOME OF THE HORMONES INVOLVED ARE:
1. 17-ketosteroids
2. Epinephrine
3. Progesterone
4. 11-oxysteroids

Ref. 1 - pp. 448-449

CHAPTER II - METABOLISM
SECTION III - LIPIDS

EACH OF THE QUESTIONS OR INCOMPLETE STATEMENTS BELOW IS FOLLOWED BY FIVE SUGGESTED ANSWERS OR COMPLETIONS. SELECT THE ONE WHICH IS BEST IN EACH CASE:

512. THE BASIC STRUCTURAL UNIT FROM WHICH ALL FATTY ACIDS ARE DERIVED IS:
A. Succinyl-CoA
B. Mevalonolactone
C. Acetoacetate
D. HMG-CoA
E. Acetyl-CoA
Ref. 5 - pp. 316-318

513. WHICH OF THE FOLLOWING COMPOUNDS IS NOT AN INTERMEDIATE IN THE BIOSYNTHESIS OF CHOLESTEROL ?:
A. Acetyl-CoA
B. Cholyl-CoA
C. Acetoacetyl-CoA
D. HMG-CoA
E. Squalene
Ref. 5 - p. 596

514. A CERTAIN COMPOUND STIMULATES THE TRANSFER OF ACETATE FROM CYTOSOL INTO MITOCHONDRIA. THIS COMPOUND IS:
A. Acetyl-CoA
B. Carnitine
C. HMG-CoA
D. TPP
E. Tetrahydrofolate
Ref. 5 - p. 228

515. IN MAN, THE SERUM LIPOPROTEIN WITH THE HIGHEST CHOLESTEROL CONTENT IS:
A. α-Lipoprotein
B. β-Lipoprotein
C. Pre-β-lipoprotein
D. Chylomicrons
E. Pre-albumin
Ref. 2 - p. 333

516. ARRANGE THE FOLLOWING ENZYMES OF CHOLESTEROL BIOSYNTHESIS IN THEIR CORRECT SEQUENCE (ACETYL-CoA → CHOLESTEROL):
A. Geranyl pyrophosphate synthetase, isopentenyl pyrophosphate synthetase, HMG-CoA reductase, squalene oxidocyclase
B. HMG-CoA condensing enzyme, mevalonic kinase, isopentenyl pyrophosphate synthetase, farnesyl pyrophosphate synthetase, squalene oxidocylcase
C. Acetoacetyl kinase, HMG-CoA condensing enzyme, phosphomevalonic kinase, mevalonic kinase
D. Farnesyl pyrophosphate synthetase, geranyl pyrophosphate synthetase, squalene synthetase, squalene oxidocyclase
E. HMG-CoA reductase, HMG-CoA condensing enzyme, mevalonic kinase, isopentenyl pyrophosphate isomerase
Ref. 2 - p. 326

517. TRIGLYCERIDES PRESENT IN PLASMA LIPOPROTEINS CAN BE HYDROLYZED TO FREE FATTY ACIDS BY THE ENZYME:
A. Lipoprotein lipase
B. Hormone-sensitive lipase
C. Heparin
D. Pancreatic lipase
E. Adenyl cyclase
Ref. 5 - pp. 342-343

518. BROWN ADIPOSE TISSUE:
A. Is present in greater amounts in adults than in children
B. Characteristically operates near maximal metabolic efficiency
C. Usually exhibits tight coupling between oxidation and phosphorylation
D. Has a higher concentration of mitochondria than white adipose tissue
E. Is usually absent in hibernators Ref. 5 - pp. 345-346

519. THE ODOR OF RANCID FATS IS DUE TO:
A. Acrolein
B. Volatile fatty acid
C. Presence of excessive amounts of Vitamin E
D. Phenol
E. Antioxidants Ref. 2 - p. 277

520. THE REICHERT-MEISSL NUMBER OF A FAT MEASURES:
A. The amount of hydroxyl groups
B. The amount of carboxyl groups
C. The unsaturation
D. The amount of volatile fatty acids
E. The amount of essential fatty acid Ref. 2 - p. 278

521. THE DEPOT FAT OF MAMMALIAN TISSUES CONSISTS LARGELY OF:
A. Glycolipids
B. Cholesterol esters
C. Cholesterol
D. Phospholipids
E. Triglycerides
Ref. 2 - pp. 269-270

522. IN THE TISSUES OF STARVED ANIMALS THE LIPIDS CONSIST LARGELY OF:
A. Glycolipids
B. Cholesterol esters
C. Cholesterol
D. Phospholipids
E. Triglycerides
Ref. 2 - p. 300-301

523. CURRENT CONCEPTS CONCERNING THE INTESTINAL ABSORPTION OF TRIGLYCERIDES ARE THAT:
A. Triglycerides must be completely hydrolyzed before the constituent fatty acids can be absorbed
B. Triglycerides are hydrolyzed partially, and the material absorbed consists of free fatty acids, mono- and diglycerides and unchanged triglyceride
C. Fatty acids with less than 10 carbon atoms are absorbed about equally via lymph and via portal blood
D. In the absence of bile the hydrolysis of triglycerides is abolished
E. Fat absorption appears to be independent of the metabolic activity of the intestinal mucosa Ref. 1 - pp. 472-473

524. THE MAJOR SITE OF ACETOACETATE FORMATION FROM FATTY ACID IS THE:
A. Liver
B. Adipose tissue
C. Intestinal mucosa
D. Kidney
E. Blood Ref. 2 - pp. 307-308

525. PANCREATIC LIPASE IS AN ENZYME WHICH:
A. Hydrolyzes triglycerides to glycerol plus fatty acid
B. Specifically attacks the 2-position of triglycerides
C. Specifically attacks the 1, 3-ester linkages of triglycerides
D. Specifically hydrolyzes 2-monoglycerides to glycerol and fatty acid
E. Specifically hydrolyzes phospholipids
Ref. 2 - p. 294

526. LECITHINASE A REMOVES A FATTY ACID MOIETY FROM LECITHIN TO FORM:
A. Phosphatidic acid
B. Inorganic phosphate
C. Choline
D. Lysolecithin
E. Glycerylphosphoryl choline
Ref. 2 - p. 324

527. A FATTY ACID NOT SYNTHESIZED IN MAN IS:
A. Oleic acid
B. Linoleic acid
C. Palmitoleic acid
D. Stearic acid
E. Palmitic acid
Ref. 2 - p. 279

528. PHOSPHOLIPIDS ARE IMPORTANT CONSTITUENTS OF CELL MEMBRANES BECAUSE:
A. They are amphoteric molecules
B. The phospholipid molecule contains a polar and a non-polar portion
C. They can form micellar solutions in water
D. They contain glycerol
E. They combine covalently with protein
Ref. 5 - pp. 179-183

529. PANCREATIC LIPASE IS A(N):
A. Oxidoreductase
B. Phospholipase
C. Dehydrogenase
D. Racemase
E. Esterase
Ref. 2 - p. 294

530. CHOLESTEROL IN LYMPH IS PRESENT PREDOMINANTLY IN THE ESTERIFIED FORM. THIS IS DUE TO THE FACT THAT:
A. Cholesterol esters are not hydrolyzed in the lumen of the intestine and are absorbed intact
B. Cholesterol esters are hydrolyzed in the lumen but are re-esterified by the intestinal mucosa
C. Cholesterol esters are hydrolyzed by the mucosa and are re-esterified in lymph
D. About 70% of serum cholesterol is present in the form of cholesterol ester
E. Cholesterol in the lumen is largely converted to coprosterol. Lymph cholesterol represents mainly newly synthesized material
Ref. 1 - pp. 474, 517, 520

531. THE "FREE FATTY ACIDS" (FFA) OF PLASMA:
A. Are metabolically inert
B. Are mainly bound to beta-liproproteins
C. Are stored in the fat depots
D. Are mainly bound to serum albumin
E. Are independent of epinephrine secretion
Ref. 1 - pp. 476, 500-503

532. IN THE CONVERSION OF CHOLESTEROL TO BILE ACIDS:
A. Side-chain oxidation precedes ring hydroxylation
B. 3-alpha, 7-alpha, 12-alpha-trihydroxy-5-alpha-cholestane is an intermediate
C. Ring hydroxylation precedes side-chain oxidation
D. Conjugation with glycine or taurine is the first step
E. Coprostanol is an intermediate
Ref. 1 - p. 522

533. KETOSIS IS ASCRIBED IN PART TO:
A. A slowdown in fat metabolism
B. An overproduction of acetyl-CoA
C. An underproduction of acetyl-CoA
D. An overutilization of glucose
E. The ketogenicity of pyruvate in the liver
Ref. 1 - pp. 503-506

534. CHYLOMICRONS CONTAIN A SMALL AMOUNT OF PROTEIN. THE ELECTROPHORETIC MOBILITY OF THIS PROTEIN CORRESPONDS TO THAT OF:
A. $Beta_2$-globulin
B. Gamma-globulin
C. $Alpha_2$-globulin
D. $Alpha_1$-globulin
E. $Beta_1$-globulin
Ref. 1 - p. 475

535. A CERTAIN SUGAR DERIVATIVE PRODUCES POLYPHAGIA AND OBESITY WHEN ADMINISTERED TO ANIMALS. THIS SUGAR IS:
A. Lipopolysaccharide
B. 5-Thio-α-D-glucopyranose
C. Sialic acid
D. D-glucosamine
E. Gold thioglucose
Ref. 2 - p. 178

536. IT HAS BEEN ESTIMATED THAT UNDER NORMAL DIETARY CONDITIONS:
A. Triglycerides are fully hydrolyzed before being absorbed
B. Triglycerides pass unchanged into the portal blood
C. Triglycerides enter the portal blood as mono- and diglycerides
D. 50% to 75% of dietary fat is split to free fatty acids which are absorbed as such
E. Triglyceride fatty acids enter the lymph largely in the form of cholesterol esters
Ref. 2 - p. 294

537. IN PHOSPHOLIPIDS OF THE CEPHALIN-TYPE CHOLINE IS REPLACED BY:
A. Sphingosine
B. Serine
C. Ethanolamine, serine and inositol
D. Inositol
E. Betaine
Ref. 2 - p. 286

538. THE CARBON CHAIN OF FATTY ACIDS IS SHORTENED BY TWO CARBON ATOMS AT A TIME. THIS INVOLVES SUCCESSIVE REACTIONS CATALYZED BY FOUR ENZYMES. THESE ACT IN THE FOLLOWING ORDER:
A. Acyl dehydrogenase, beta-hydroxyacyl dehydrogenase, enoyl hydrase, thiolase
B. Acyl dehydrogenase, enoyl hydrase, beta-hydroxyacyl dehydrogenase, thiolase
C. Acyl dehydrogenase, thiolase, enoyl hydrose, beta-hydroxyacyl dehydrogenase
D. Enoyl hydrase, beta-hydroxyacyl dehydrogenase, acyl dehydrogenase, thiolase
E. Acyl dehydrogenase, enoyl hydrase, beta-hydroxyacyl dehydrogenase, acyl dehydrogenase
Ref. 1 - p. 479

ANSWER THE FOLLOWING QUESTIONS BY USING THE KEY OUTLINED BELOW:
A. If A is greater than B
B. If B is greater than A
C. If A and B are equal or nearly equal

539. A. Liver lipid levels after administration of lipotropic agents
B. Liver lipid levels after administration of phosphorus
Ref. 2 - pp. 302-304

540. A. Concentration of plasma esterified cholesterol
B. Concentration of plasma free cholesterol
Ref. 1 - p. 79

541. A. Moles of glycerol per mole of lecithin
B. Moles of glycerol per mole of cerebroside
Ref. 2 - p. 289

542. A. Cholesterol content of α-liproproteins
B. Cholesterol content of β-lipoproteins
Ref. 2 - p. 333

543. A. Triglyceride content of α-lipoprotein
B. Triglyceride content of β-lipoprotein
Ref. 2 - p. 333

544. A. Serum level of α-lipoprotein in Tangier disease
B. Serum level of β-lipoprotein in acanthocytosis
Ref. 2 - p. 334

545. A. Iodine number of triolein
B. Iodine number of tripalmitin
Ref. 2 - p. 278

546. A. Acetyl number of olive oil
B. Acetyl number of castor oil
Ref. 2 - pp. 278-279

547. A. Reichert-Meissl number of butter fat
B. Reichert-Meissl number of peanut oil
Ref. 2 - p. 278

548. A. Number of carbon atoms in linoleic acid
B. Number of carbon atoms in arachidonic acid
Ref. 2 - p. 279

549. A. Saponification number of butter
B. Saponification number of oleomargarine
Ref. 2 - p. 278

550. A. Galactose content of cephalins
B. Glycerol content of cerebrosides
Ref. 2 - pp. 286-289

MATCH THE FORMULAS WITH THE CORRECT NAME:

A.

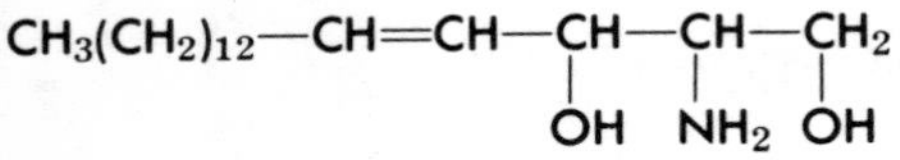

$$CH_3(CH_2)_{12}-CH{=}CH-\underset{\displaystyle OH}{\underset{|}{CH}}-\underset{\displaystyle NH_2}{\underset{|}{CH}}-\underset{\displaystyle OH}{\underset{|}{CH_2}}$$

B.

$$CH_3(CH_2)_4-CH{=}CH-CH_2-CH{=}CH-(CH_2)_7-COOH$$

C.

$$CH_3-CH_2-CH{=}CH-CH_2-CH{=}CH-CH_2-CH{=}CH-(CH_2)_7-COOH$$

D.

$$CH_3(CH_2)_4CH{=}CHCH_2CH{=}CHCH_2CH{=}CHCH_2CH{=}CH(CH_2)_3COOH$$

E.

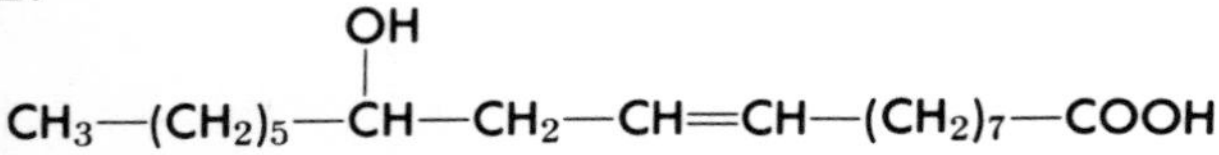

$$CH_3-(CH_2)_5-\overset{\displaystyle OH}{\overset{|}{CH}}-CH_2-CH{=}CH-(CH_2)_7-COOH$$

551. ___ Ricinoleic acid

552. ___ Linoleic acid

553. ___ Sphingosine

554. ___ Linolenic acid

555. ___ Arachidonic acid

Ref. 1 - pp. 61-62; 72

MATCH THE FORMULAS WITH THE CORRECT NAME:

A.

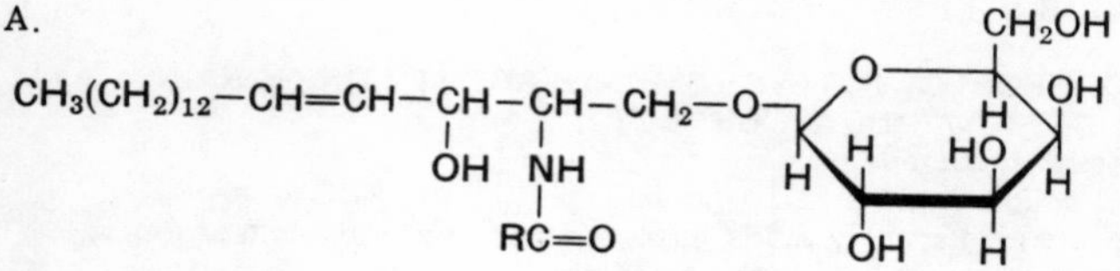

$CH_3(CH_2)_{12}-CH{=}CH-CH(OH)-CH(NHC(=O)R)-CH_2-O-$ (galactose ring: CH_2OH, OH, HO, OH, H)

B.

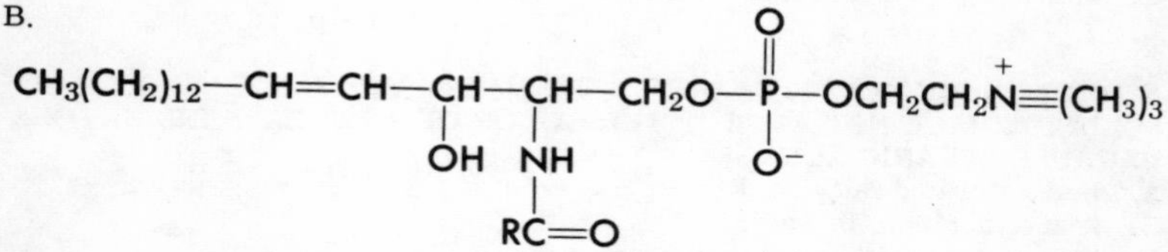

$CH_3(CH_2)_{12}-CH{=}CH-CH(OH)-CH(NHC(=O)R)-CH_2O-P(=O)(O^-)-OCH_2CH_2\overset{+}{N}{\equiv}(CH_3)_3$

C.

H_2COOCR
$R'COOCH$
$H_2C-O-P(=O)(O^-)-OCH_2CH_2\overset{+}{N}H_3$

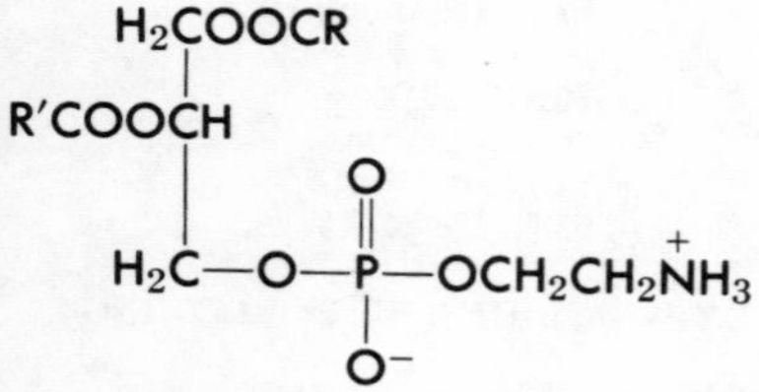

D.

$CH_3-(CH_2)_7-CH{=}CH-(CH_2)_7-COOH$

E.

$CH_3-(CH_2)_5-CH{=}CH-(CH_2)_7-COOH$

556. ___ Phosphatidyl ethanolamine

557. ___ Palmitoleic acid

558. ___ Sphingomyelin

559. ___ Oleic acid

560. ___ Cerebroside

Ref. 1 - pp. 59; 70; 73-74

EACH OF THE QUESTIONS OR INCOMPLETE STATEMENTS BELOW IS FOLLOWED BY 5 SUGGESTED ANSWERS OR COMPLETIONS. SELECT THE ONE WHICH IS BEST IN EACH CASE:

561. WHICH OF THE FOLLOWING STATEMENTS ABOUT LONG-CHAIN SATURATED FATTY ACIDS IS CORRECT?:
A. Contain ester linkages
B. All have a double bond in 9, 10-position
C. Naturally-occurring fatty acids have odd number of carbon atoms
D. Boiling point increases with chain length
E. Melting point decreases with chain length
Ref. 1 - p. 58

562. WHEN STEARIC ACID IS SHAKEN WITH WATER THE RESULTING SOLUTION WILL NOT AFFECT THE COLOR OF ACID-BASE INDICATORS BECAUSE STEARIC ACID:
A. Is a saturated fatty acid
B. Is too insoluble in water
C. Is a much weaker acid than acetic acid
D. Has a high iodine number
E. Has an odd number of carbon atoms
Ref. 1 - p. 58

563. WHICH OF THE FOLLOWING IS NOT A PHOSPHOLIPID?:
A. Cerebroside
B. Plasmalogen
C. Sphingomyelin
D. Cephalin
E. Lecithin
Ref. 1 - p. 73

564. SOAPS ARE CHARACTERIZED BEST BY WHICH OF THE FOLLOWING STATEMENTS?:
A. They are responsible for the hardness of water
B. They are anionic detergents
C. They are hydrophobic, but not hydrophilic
D. They are produced by hydrolyzing long-chain hydrocarbons
E. They increase surface tension
Ref. 1 - p. 62

565. CATALYTIC HYDROGENATION OF LINOLEIC ACID PRODUCES:
A. Linolenic acid
B. Palmitic acid
C. 9, 10-dihydroxystearic acid
D. Palmitoleic acid
E. Stearic acid
Ref. 1 - p. 63

566. 10 MG OF AN UNKNOWN MONOCARBOXYLIC FATTY ACID REQUIRED 5.0 ML OF 0.01 N NaOH WHEN TITRATED TO A PHENOLPHTHALEIN END POINT. ITS MOLECULAR WEIGHT WAS APPROXIMATELY:
A. 2
B. 500
C. 200
D. 50
E. 20
Ref. 1 - p. 69

567. THE IODINE NUMBER OF A FAT IS DEFINED AS THE NUMBER OF GRAMS OF IODINE REACTING WITH 100 G OF THAT FAT. IF THE MOLECULAR WEIGHT OF LINOLEIC ACID IS 280 AND THE ATOMIC WEIGHT OF IODINE IS 127 WHAT IS THE APPROXIMATE IODINE NUMBER OF LINOLEIC ACID?:
A. 0
B. 90
C. 135
D. 180
E. 270
Ref. 1 - p. 64

568. HYDROLYSIS OF A TRIGLYCERIDE IS ALSO KNOWN AS:
A. Saponification
B. Esterification
C. Hydrogenation
D. Dehydration
E. Rancidification
Ref. 1 - p. 67

569. IF THE CHOLINE-MOIETY OF LECITHIN IS REPLACED BY β-ETHANOLAMINE THE COMPOUND PRODUCED IS A(N):
A. Sphingomyelin
B. Plasmalogen
C. Inositide
D. Cerebroside
E. Cephalin
Ref. 2 - p. 286

570. WHICH OF THE FOLLOWING STATEMENTS ABOUT CHOLESTANOL IS INCORRECT?:
A. It is structurally related to the terpenes
B. It is optically active
C. It is unsaturated
D. It has the A/B-trans configuration
E. It is an alcohol
Ref. 1 - p. 78

571. WHICH STATEMENT EXPRESSES MOST ACCURATELY THE RELATIONSHIP BETWEEN ALPHA-CAROTENE, RETINOL AND SQUALENE?:
A. They may be thought of as vitamins
B. They have the same chemical formula
C. They are structurally related to 2-methylbutadiene
D. They are stereoisomers
E. They are geometrical isomers
Ref. 1 - pp. 75-76

572. COPROSTEROL (I) AND CHOLESTANOL (II) HAVE THE SAME CHEMICAL FORMULA. THEY DIFFER CHEMICALLY BECAUSE:
A. (I) and (II) are optical isomers
B. (I) is a natural product
C. The 5-hydrogen is beta in (I) and alpha in (II)
D. The 3-hydroxyl is alpha in (I) and beta in (II)
E. The 5-hydrogen is alpha in (I) and beta in (II)
Ref. 1 - p. 79

573. DEOXYCHOLIC ACID AND CHENODEOXYCHOLIC ACID HAVE THE SAME CHEMICAL FORMULA. THEY DIFFER CHEMICALLY BECAUSE:
A. They are positional isomers
B. They are geometrical isomers
C. They are stereoisomers
D. They differ in the position of the double bond
E. They differ in the attachment of rings A and B
Ref. 1 - p. 81

574. A MAJOR STRUCTURAL DIFFERENCE BETWEEN ESTROGENS AND ANDROGENS IS THE FACT THAT:
A. The androgens are usually C_{21} steroids
B. The estrogens are usually digitonin-precipitable
C. The androgens have an aromatic A ring
D. The estrogens have an aromatic A ring
E. The androgens have phenolic hydroxyls
Ref. 1 - p. 82

575. THE ANGULAR METHYL GROUP AT THE 10-POSITION OF STEROIDS:
A. Has been transformed into a methylene group in ergosterol
B. Reacts readily with Girard's reagent
C. Is absent in progesterone derivatives
D. Is chemically highly reactive
E. Is a reference point for describing the orientation of 5-substituents
Ref. 1 - pp. 78, 79

576. IN MAMMALIAN BILE, THE BILE ACIDS ARE USUALLY PRESENT:
A. In the free, uncombined form
B. Conjugated with a variety of amino acids
C. In the form of cholesterol esters
D. Conjugated with glycine or taurine
E. Conjugated with water-soluble amides
Ref. 1 - p. 81

577. IN STARVATION ALL OF THE FOLLOWING ENZYME ACTIVITIES ARE DECREASED, EXCEPT:
A. Lipoprotein lipase
B. Citrate cleavage enzyme
C. Fatty acid synthase
D. Acetyl-CoA carboxylase
E. CoA-carnitine acyl transferase Ref. 5 - p. 549

EACH GROUP OF QUESTIONS BELOW CONSISTS OF FIVE LETTERED HEADINGS FOLLOWED BY A LIST OF NUMBERED WORDS OR PHRASES. FOR EACH NUMBERED WORD OR PHRASE SELECT THE ONE LETTERED HEADING THAT IS MOST CLOSELY RELATED TO IT:

A. Neutral fats
B. Phosphatides
C. Cerebrosides
D. Waxes
E. Terpenes

578. ___ Hydrolysis yields three molecules of fatty acid and one molecule of glycerol
579. ___ Fatty acid esters of alcohols other than glycerol
580. ___ Contain isoprene-like fragments
581. ___ Sometimes referred to as galactolipids
582. ___ May contain choline, ethanolamine or serine
Ref. 1 - pp. 66, 69, 73-75

IN THE FREDRICKSON CLASSIFICATION FIVE TYPES OF PRIMARY HYPERLIPOPROTEINEMIA ARE RECOGNIZED. MATCH THE HYPERLIPOPROTEINEMIA WITH THE CORRECT LIPOPROTEIN WHICH IS ALTERED OR ELEVATED:

A. Type I
B. Type II
C. Type III
D. Type IV
E. Type V

583. ___ Pre-β-lipoprotein
584. ___ Pre-β-lipoprotein and chylomicrons
585. ___ β-Lipoprotein
586. ___ Chylomicrons
587. ___ Abnormal β-lipoprotein Ref. 2 - p. 334

A. Glycerol, fatty acid, fatty aldehyde, phosphoric acid, choline
B. Glycerol, fatty acid, phosphoric acid
C. Glycerol, fatty acid, choline, phosphoric acid
D. Spingosine, fatty acid, galactose
E. Sphingosine, fatty acid, phosphoric acid, choline

588. ___ Plasmalogen
589. ___ Lecithin
590. ___ Phosphatidic acid
591. ___ Sphingomyelin
592. ___ Cerebroside Ref. 2 - pp. 285-289

A. Sphingosine
B. Lecithin
C. Phosphatidyl ethanolamine
D. Phosphatidyl inositol
E. S-Adenosylmethionine

593. ___ Synthesized from cytidine triphosphate, phosphoethanolamine and diglyceride
594. ___ Synthesized from palmitic aldehyde + serine
595. ___ Synthesized from diglyceride + CDP-choline
596. ___ Required for de novo lecithin synthesis
597. ___ Synthesis occurs from CDP-diglyceride but does not involve phosphatidyl serine Ref. 1 - pp. 509-514

A. Plasmalogens
B. Cardiolipins
C. Ceramides
D. Lecithin, cephalin
E. Gangliosides

598. ___ Characterized by an α, β-unsaturated ether linkage
599. ___ Diphosphatidyl glycerols
600. ___ N-acylsphingosines
601. ___ Contain sialic acid
602. ___ Derivatives of phosphatidic acid Ref. 1 - pp. 71-74
Ref. 2 - pp. 285-289

ANSWER THE FOLLOWING QUESTIONS BY USING THE KEY OUTLINED BELOW:
A. If 1, 2 and 3 are correct
B. If 1 and 3 are correct
C. If 2 and 4 are correct
D. If only 4 is correct
E. If all four are correct

603.
1. Singly unsaturated fatty acids occur in nature exclusively in the trans (rather than cis) configuration
2. Multiple unsaturation in naturally occurring fatty acids is generally of the conjugated type
3. The most abundant polyunsaturated fatty acids are oleic and stearic acids
4. Double bonds in unsaturated fatty acids occur primarily after the ninth carbon atom Ref. 1 - p. 59

604.
1. Salts of fatty acids are termed "soaps"
2. Salts of fatty acids and alkali metals are usually insoluble in water
3. Opalescence of solutions of salts of fatty acids indicates the presence of polymolecular micelles
4. A high iodine number is indicative of complete saturation of a fatty acid Ref. 1 - p. 62

605.
1. Even-numbered fatty acids of less than 10 carbon atoms are all liquids at room temperature
2. The most abundant saturated fatty acid in animal lipids is palmitic acid
3. Milk fat contains appreciable concentrations of low molecular weight fatty acids
4. Unsaturation of fatty acids depresses the melting point and increases alcohol solubility Ref. 1 - pp. 58-61

606.
1. The double bond in naturally-occurring singly unsaturated fatty acids is in the 2-3 position
2. The common unsaturated fatty acids are solids at room temperature
3. The most abundant fatty acid in nature is arachidonic acid
4. The most abundant singly unsaturated fatty acids are oleic and palmitoleic acids Ref. 1 - p. 59

607. THE ROLE OF CARNITINE IN CELLULAR METABOLISM:
1. Essential for biosynthesis of fatty acids with an odd number of carbon atoms
2. Essential for extracellular transport of activated fatty acids
3. Catalyzes the cyclization of squalene
4. Essential for the intracellular transport of activated fatty acids

Ref. 5 - p. 228

608. REGARDING DISEASES OF LIPID METABOLISM IN MAN:
1. Tangier disease is due to a hereditary deficiency of α-lipoprotein
2. Tay-Sachs disease can frequently be diagnosed by the presence of tendon xanthoma
3. In acanthocytosis there is a deficiency of serum β-lipoproteins
4. Type II hyperlipoproteinemia is associated with an increase in serum triglyceride level when carbohydrate is fed

Ref. 2 - pp. 333-336

MATCH THE FORMULAS WITH THE CORRECT NAME:

A.

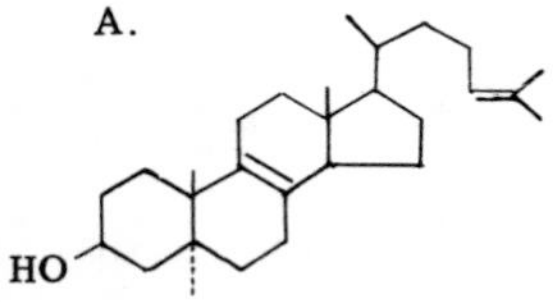

HO

B.

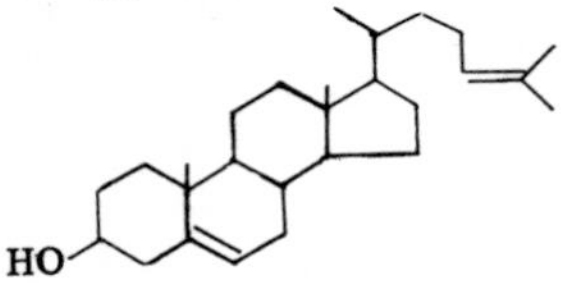

HO

C.

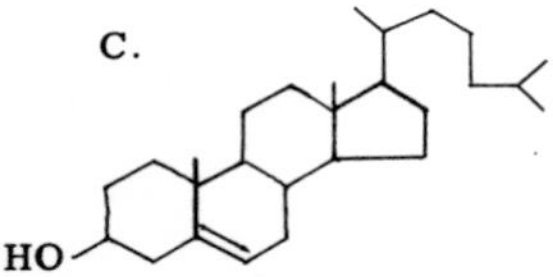

HO

D.

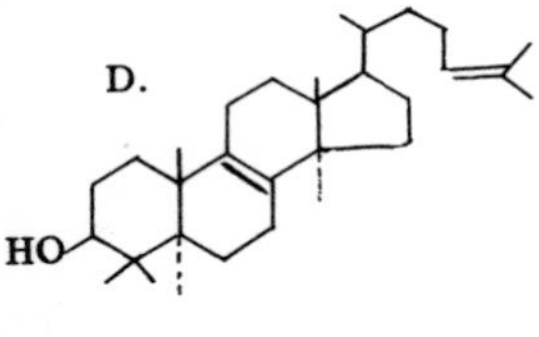

HO

E.

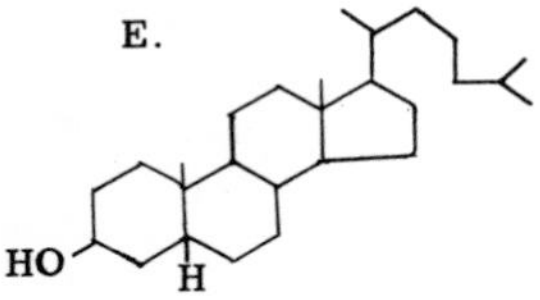

HO H

609. ___ Lanosterol
610. ___ Cholesterol
611. ___ Desmosterol
612. ___ Zymosterol
613. ___ Coprostanol

Ref. 1 - pp. 521-522

PHOSPHATIDES ARE DEGRADED BY SPECIFIC HYDROLYTIC ENZYMES. MATCH THE ENZYMES TO THE BONDS THEY ATTACK:

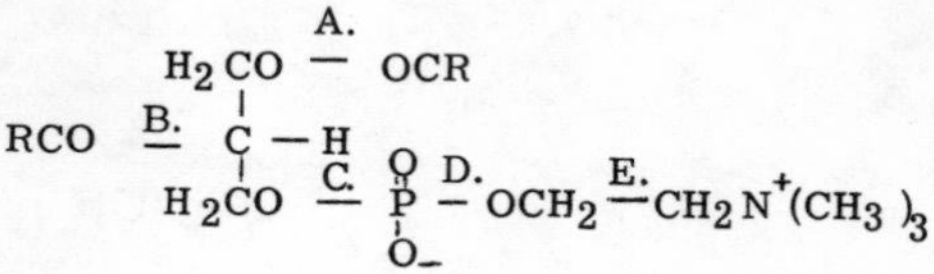

$$\begin{array}{l} \quad\quad\quad H_2CO \overset{A.}{—} OCR \\ RCO \overset{B.}{—} C — H \\ \quad\quad\quad H_2CO \overset{C.}{—} \overset{O}{\underset{O^-}{P}} \overset{D.}{—} OCH_2 \overset{E.}{—} CH_2N^+(CH_3)_3 \end{array}$$

614. ___ Phospholipase A
615. ___ Lysophospholipase
616. ___ Phospholipase D
617. ___ Phospholipase C
618. ___ Not known

Ref. 1 - p. 513

MATCH THE FORMULAS WITH THE CORRECT NAME

A. [structure with OH, COOH, HO, H, OH]

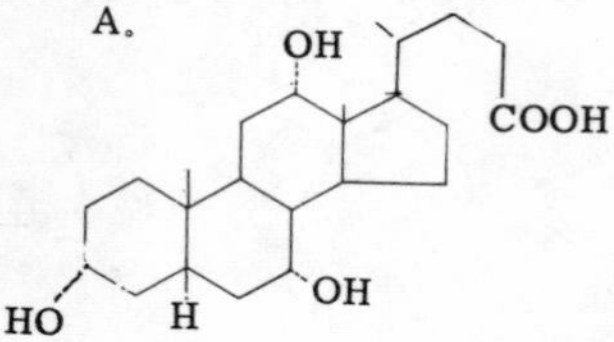

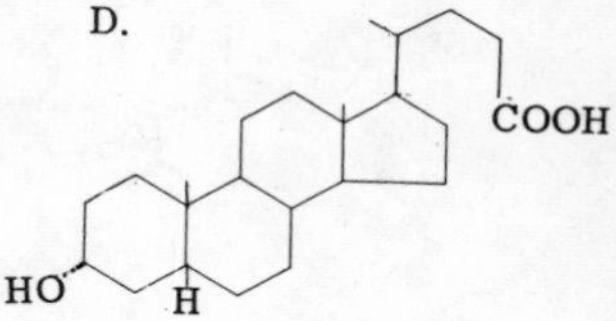

B. [structure with COOH, HO, H, OH]

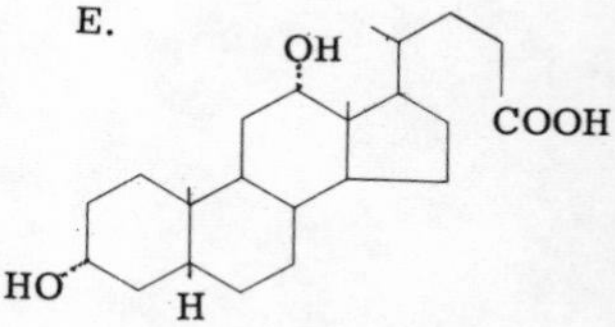

C. [structure with COOH, H]

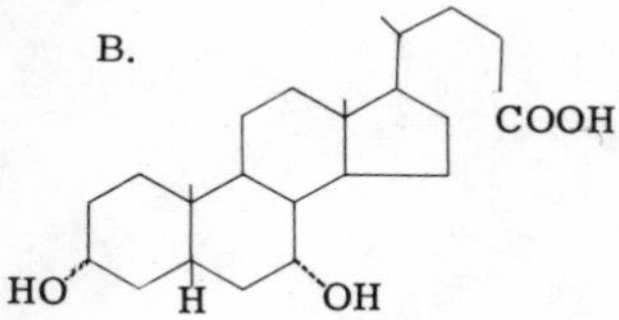

D. [structure with COOH, HO, H]

E. [structure with OH, COOH, HO, H]

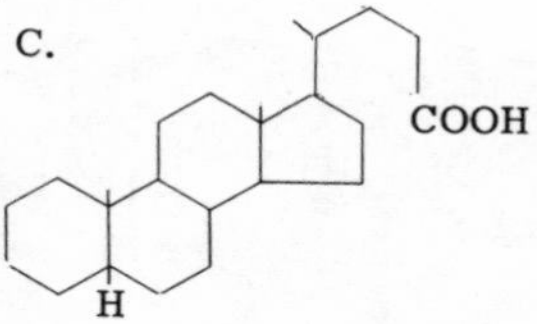

619. ___ Cholanic (cholanoic) acid
620. ___ Chenodeoxycholic acid
621. ___ Deoxycholic acid
622. ___ Lithocholic acid
623. ___ Cholic acid

Ref. 1 - p. 81

RELATE THE STRUCTURES OF THE FOLLOWING TERPENES TO THEIR COMMON NAMES:

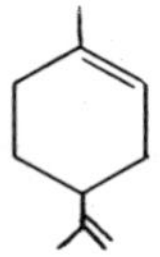

OH

A.

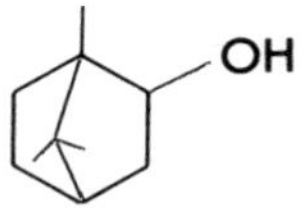

B.

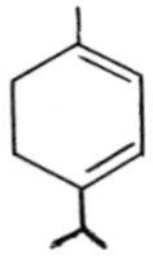

C.

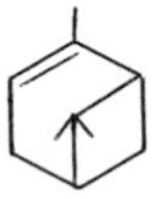

D.

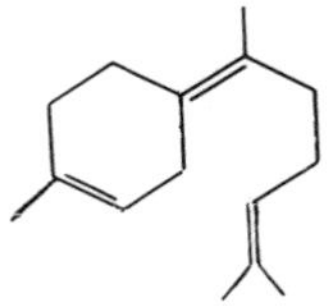

E.

624. ___ Borneol
625. ___ α -Pinene
626. ___ α -Terpinene
627. ___ Bisabolene
628. ___ Limonene

Ref. 3 - pp. 752-753

MATCH THE DISEASE STATE WITH THE CORRECT ENZYME DEFICIENCY ASSOCIATED WITH IT:

A. Sphingomyelinase
B. Aryl sulfatase A
C. Ceramide trihexosidase
D. Glucocerebrosidase
E. N-acetyl gelactosaminidase

629. ___ Gaucher's disease
630. ___ Metachromatic leukodystrophy
631. ___ Niemann-Pick disease
632. ___ Fabry's disease
633. ___ Tay-Sachs disease

Ref. 2 - p. 336

A. Mevalonic acid
B. Farnesyl pyrophosphate
C. Zymosterol
D. Lanosterol
E. Ergosterol

634. ___ Source of isoprenoid unit in cholesterol synthesis
635. ___ A precursor of squalene which is derived from geranyl pyrophosphate
636. ___ Synthesized by action of squalene oxide cyclase upon squalene
637. ___ A precursor of cholesterol which arises by removal of three methyl groups from lanosterol
638. ___ Methylated by S-adenosyl-methionine

Ref. 1 - pp. 517-521

A. Coprostanol
B. Δ^5-pregnenolone
C. Bile acid
D. Glycine
E. Cholyl-CoA

639. ___ Product of microbial reduction of cholesterol in the intestine
640. ___ Main metabolic end product of cholesterol
641. ___ A precursor of progesterone which arises from splitting of the side chain of cholesterol between C-20 and C-22
642. ___ Intermediate in bile acid synthesis
643. ___ Forms conjugates with CoA derivatives of bile acids

Ref. 1 - pp. 522-524

A. Tangier disease
B. Abetalipoproteinemia
C. Metachromatic leukodystrophy
D. Gaucher's disease
E. Niemann-Pick disease

644. ___ Accumulation of sulfatides
645. ___ Accumulation of glucocerebrosides
646. ___ Absence of plasma high density lipoproteins
647. ___ Accumulation of sphingomyelin
648. ___ Absence of plasma low density lipoproteins

Ref. 1 - pp. 527-528

A. Acyl dehydrogenase
B. Enoyl hydrase
C. Beta-hydroxyacyl dehydrogenase
D. Thiolase
E. Thiokinase

649. ___ Catalyzes formation of CoA derivatives from fatty acid, CoA and ATP
650. ___ Catalyzes reversible hydration of unsaturated fatty acyl CoA
651. ___ FAD-containing enzyme, catalyzing formation of alpha, beta-unsaturated fatty acyl CoA derivative
652. ___ DPN-requiring enzyme, catalyzing conversion of beta-hydroxyacyl CoA to beta-ketoacyl CoA
653. ___ Splits beta-ketoacyl CoA $\longrightarrow$ acetyl-CoA + fatty acyl CoA with 2 fewer carbon atoms

Ref. 1 - pp. 480-481
Ref. 3 - p. 594

A. Crotonase
B. Fatty acid spiral
C. Acyl carrier protein
D. Lipoprotein lipase
E. CDP-choline

654. ___ β-Oxidation
655. ___ Fatty acid synthesis
656. ___ Acyl-CoA hydrase
657. ___ Lecithin synthesis
658. ___ Clearing factor

Ref. 2 - pp. 298, 307, 308, 311

A. Acetyl-CoA
B. Coenzyme A
C. Propionyl CoA
D. HMG-CoA
E. Phenylacetic acid

659. ___ Forms thiol esters with fatty acids
660. ___ Oxidation product of fatty acid with odd number of carbon atoms
661. ___ Links fatty acid metabolism to citric acid cycle
662. ___ Product of oxidation of phenyl derivatives of fatty acids with even number of carbon atoms
663. ___ Intermediate in acetoacetate synthesis

Ref. 1 - pp. 480-483, 503-504
Ref. 2 - pp. 304, 310, 315

A. Dietary choline
B. Arsenocholine
C. Serotonin
D. S-Adenosylmethionine
E. Guanidoacetic acid

664. ___ Promotes lipid mobilization from adipose tissue
665. ___ Is a controlling factor in the rate of lecithin synthesis in the liver
666. ___ Prevents fatty liver in absence of both exogenous and endogenous choline
667. ___ A source of methyl groups for choline synthesis
668. ___ Will enhance fatty liver by diverting methyl groups from choline synthesis Ref. 1 - pp. 500-503

ANSWER THE FOLLOWING QUESTIONS BY USING THE KEY OUTLINED BELOW:
A. If A is greater than B
B. If B is greater than A
C. If A and B are equal or nearly equal

669. A. Concentration of blood lecithin
B. Concentration of blood sphingomyelin
Ref. 1 - pp. 475, 706

670. A. Concentration of blood phosphatidyl ethanolamine
B. Concentration of blood lecithin Ref. 1 - pp. 475, 706

671. A. Free lipid in blood plasma
B. Lipid associated with protein in blood plasma
Ref. 1 - pp. 475-476, 706

672. A. Efficiency of de novo fatty acid synthesis in isolated mitochondria
B. Efficiency of de novo fatty acid synthesis in high-speed supernatant fraction of cell (non-mitochondrial) Ref. 1 - pp. 486; 490

673. A. Net synthesis of carbohydrate from fatty acids in mammals
B. Net synthesis of carbohydrate from fatty acids in germinating seeds
Ref. 1 - pp. 413; 498-499

674. A. Weight of one caloric equivalent of lipid
B. Weight of one caloric equivalent of polysaccharide
Ref. 1 - p. 477

675. A. Degree of hydration of depot polysaccharide
B. Degree of hydration of depot lipid Ref. 1 - p. 477

676. A. Number of double bonds in oleic acid
B. Number of double bonds in vaccenic acid
Ref. 1 - p. 494

677. A. Lipid content of alpha-lipoprotein
B. Lipid content of beta-lipoprotein Ref. 1 - p. 475

678. A. Molecular weight of high-density lipoprotein
B. Molecular weight of low-density lipoprotein
Ref. 1 - p. 711

679. A. Unesterified fatty acid bound to globulin
B. Unesterified fatty acid bound to albumin
Ref. 1 - p. 710

680. A. Glycine-conjugated bile acid in human bile
B. Taurine-conjugated bile acid in human bile
Ref. 1 - p. 820

681. A. De novo biosynthesis of fatty acids in mitochondria
B. De novo biosynthesis of fatty acids in cytosol
Ref. 1 - p. 492

682. A. Binding of fatty acid by acyl carrier protein
B. Binding of fatty acid-CoA by acyl carrier protein
Ref. 1 - p. 486
Ref. 3 - pp. 719-720

683. A. Serum cholesterol levels in Tangier disease
B. Serum cholesterol levels in xanthomatosis
Ref. 1 - pp. 527-528

684. A. Formation of 7α-hydroxycholesterol from cholesterol in liver
B. Formation of coprostanol from cholesterol in liver
Ref. 1 - pp. 522-523

ANSWER THE FOLLOWING QUESTIONS BY USING THE KEY OUTLINE BELOW:
A. If 1, 2 and 3 are correct
B. If 1 and 3 are correct
C. If 2 and 4 are correct
D. If only 4 is correct
E. If all four are correct

685. 1. In normal mammals about 10% of the body weight may be lipid, almost all of which consists of phosphatides and sterols
2. The lipid content of the body is uniformly distributed throughout all cells of the body
3. Lipids yield about 1/2 the calories per gram as compared to carbohydrate
4. Lipids are stored in a relatively water-free state in tissues
Ref. 1 - pp. 476-478

686. 1. The more saturated a lipid the larger the energy yield on complete oxidation
2. When depleted depot fat is restored by a lipid-rich diet the physical and chemical character of the new depot fat resembles that of the dietary lipid
3. The depot lipid stores of the body are in a dynamic steady state
4. Depot lipids of man consist largely of triglyceride
Ref. 1 - pp. 477-479, 495

687. WHICH OF THE FOLLOWING REACTIONS ARE INVOLVED IN FATTY ACID OXIDATION?:
1. Dehydrogenation of fatty acid CoA derivative by an FAD-containing enzyme to alpha, beta-unsaturated derivative
2. Hydration of the double bond to form the beta-hydroxy compound
3. DPN - dependent dehydrogenation to yield the beta-keto derivative
4. Reaction of the beta-keto derivative with an enzyme which splits off acetic acid to yield a fatty acid of two fewer carbons

Ref. 1 - p. 479

688. INTESTINAL ABSORPTION OF FATS:
1. Is aided by bile salts
2. Involves the absorption of intact neutral fats only
3. Involves glycerides as well as free fatty acids and glycerol
4. Is independent of enzymatic activity of the intestinal wall

Ref. 1 - pp. 473-474

689.
1. Fatty acid metabolism is closely linked to carbohydrate metabolism
2. Lipogenesis is coupled to glucose-6-phosphate oxidation
3. Fatty acid oxidation is influenced by glycolysis
4. TPNH required for fatty acid synthesis may be generated by action of transhydrogenase

Ref. 1 - pp. 482; 485; 497-499

690. UNESTERIFIED FATTY ACIDS APPEAR IN PLASMA FROM FAT DEPOTS BECAUSE OF:
1. Mobilization of depot lipid
2. Hydrolysis of chylomicra by clearing factor
3. Hydrolysis of plasma lipoproteins by clearing factor
4. Mobilization of liver lipid

Ref. 1 - pp. 500-503

691. CHOLESTEROL HAS A NUMBER OF IMPORTANT PHYSIOLOGICAL FUNCTIONS WHICH REQUIRE ITS TRANSFORMATION TO OTHER SUBSTANCES. THESE INCLUDE:
1. Aromatization of ring A to form adrenal steroids
2. Scission of the side chain between C-20 and C-22, leading eventually to the formation of progesterone
3. Scission of the side chain between C-19 and C-20 to produce androgenic steroids
4. Formation of 7-alpha-hydroxycholesterol which is the precursor of the bile acids

Ref. 1 - pp. 522-524;939;946

692. PHOSPHATIDES:
1. Occur in membranes of erythrocytes
2. Stabilize chylomicra
3. Participate in blood clotting
4. Occur in myelin sheath of nerves

Ref. 1 - pp. 507; 735

693.
1. Absorption of cholesterol is dependent upon the presence of bile salts in the intestinal lumen
2. During the process of absorption the major portion of cholesterol is esterified with fatty acids
3. All the carbon atoms of endogenous cholesterol are derived from acetyl CoA
4. Cholesterol synthesis is catalyzed by enzymes occurring predominantly in the supernatant fluid obtained from sedimentation of microsomes

Ref. 1 - pp. 474, 517

694. FATTY LIVER IN UNTREATED DIABETES RESULTS FROM:
1. Excessive fatty acid synthesis in liver
2. Loss of ability of liver to oxidize fatty acids
3. Decreased cholesterol biosynthesis in liver
4. Excessive migration of depot lipid to the liver

Ref. 1 - pp. 506, 525

695. IN MAMMALIAN FATTY ACID BIOSYNTHESIS:
1. The mitochondrial fatty acid synthesizing system serves mainly as a chain-elongation system
2. The soluble system requires biotin as a cofactor which forms a complex with CO_2. This explains why in the presence of $C^{14}O_2$ the fatty acids formed contain C^{14}
3. The reaction stops at C-16 or C-18 fatty acids as the result of the hydrolytic action of a deacylase
4. The pathway is the same as that of fatty acid degradation since only CoA derivatives of the fatty acids can participate

Ref. 1 - pp. 485-486, 489-490, 492, 1033-1034
Ref. 2 - p. 311

MATCH THE FORMULA WITH THE CORRECT NAME:

A.

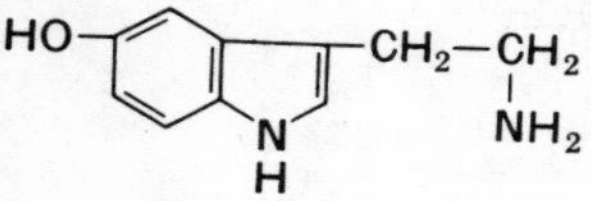

HO— $-CH_2-CH_2$ | NH_2
N
H

B.

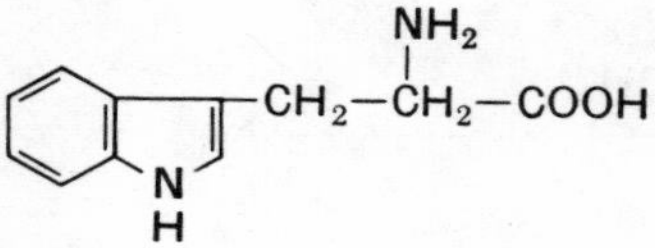

NH_2 | $-CH_2-CH_2-COOH$
N
H

C.

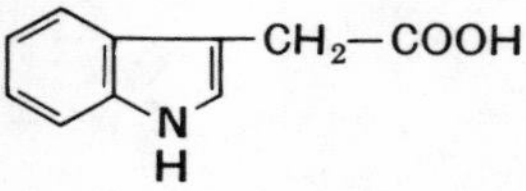

$-CH_2-COOH$
N
H

D.

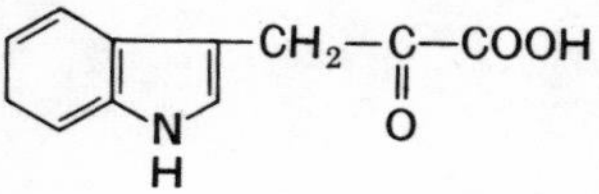

$-CH_2-C-COOH$ ‖ O
N
H

E.

NH_2 | $-CH_2-CH_2$
N
H

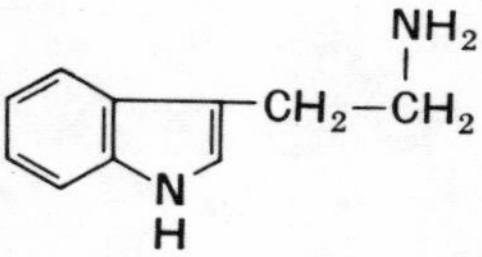

696. ___ Serotonin

697. ___ Tryptophan

698. ___ Tryptamine

699. ___ Indoleacetic acid

700. ___ Indolepyruvic acid

Ref. 1 - p. 590

A.

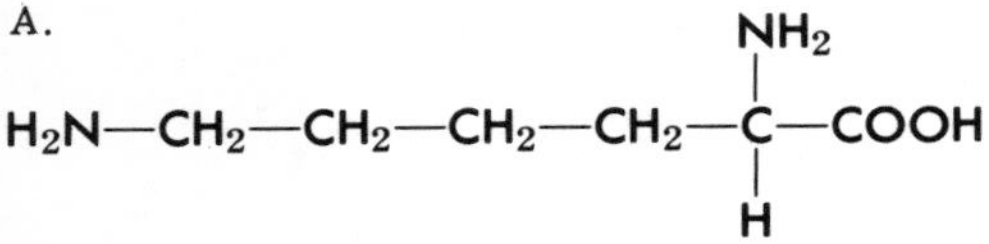

$$H_2N{-}CH_2{-}CH_2{-}CH_2{-}CH_2{-}\overset{NH_2}{\underset{H}{C}}{-}COOH$$

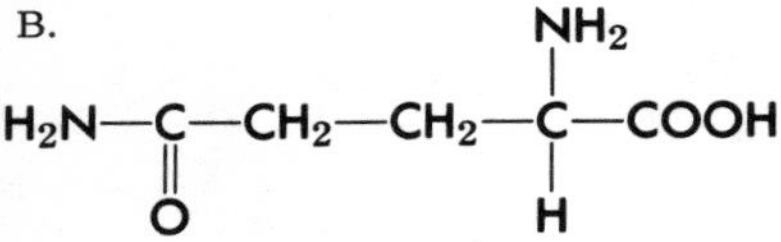

B.

$$H_2N{-}\underset{\overset{\|}{O}}{C}{-}CH_2{-}CH_2{-}\overset{NH_2}{\underset{H}{C}}{-}COOH$$

C.

$$HOOC{-}CH_2{-}\overset{NH_2}{\underset{H}{C}}{-}COOH$$

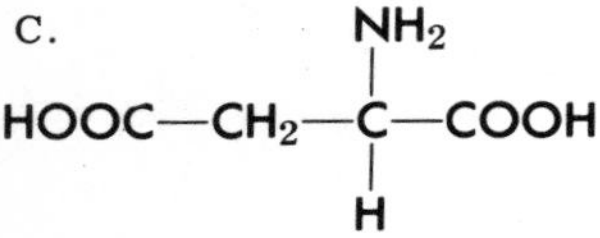

D.

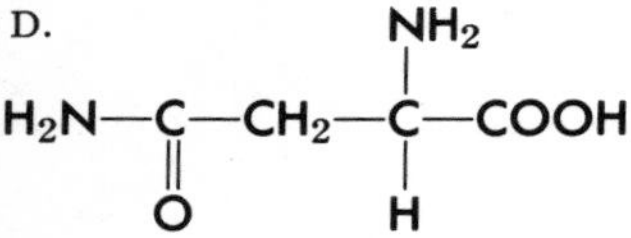

$$H_2N{-}\underset{\overset{\|}{O}}{C}{-}CH_2{-}\overset{NH_2}{\underset{H}{C}}{-}COOH$$

E.

$$HO{-}CH_2{-}\overset{NH_2}{\underset{H}{C}}{-}COOH$$

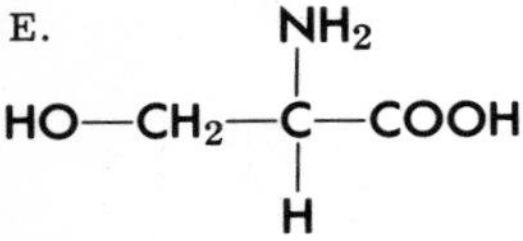

701. ___ Lysine

702. ___ Glutamine

703. ___ Asparagine

704. ___ Aspartic acid

705. ___ Serine

Ref. 1 - pp. 89; 91-93

THE BIOLOGICAL METHYLATING AGENT S-ADENOSYLMETHIONINE IS ACTIVE WITH A VARIETY OF SUBSTRATES. MATCH THE SUBSTRATES OF THE BIOLOGICAL METHYLATION WITH THE PRODUCTS LISTED BELOW:

A. N-acetyl-5-hydroxytryptamine
B. γ-Aminobutyric acid
C. Carnosine
D. Guanidoacetic acid
E. Phosphatidylethanolamine

706. ___ Melatonin
707. ___ Creatine
708. ___ γ-Butyrobetaine
709. ___ Lecithin
710. ___ Anserine Ref. 1 - p. 583

EACH OF THE QUESTIONS OR INCOMPLETE STATEMENTS BELOW IS FOLLOWED BY FIVE SUGGESTED ANSWERS OR COMPLETIONS. SELECT THE ONE WHICH IS BEST IN EACH CASE:

711. PROTEINS ARE CHARACTERIZED BY:
A. Rapid diffusion rates compared to salts
B. Behavior as amphoteric dipolar ions
C. Lack of specific molecular configuration
D. Great stability under various physical conditions
E. Electrophoretic immobility Ref. 1 - pp. 118-121

712. WHEN A PROTEIN IS HYDROLYZED THERE IS:
A. A decrease in free carboxyl groups
B. An increase in free amino groups
C. A large decrease in pH
D. Formation of peptide bonds
E. A decrease in free amino groups Ref. 1 - p. 88

713. ANSERINE AND CARNOSINE ARE:
A. Essential amino acids
B. Tripeptides
C. Enzymes
D. Histones
E. Dipeptides
Ref. 1 - p. 112

714. A DIPEPTIDE, WHICH CONTAINS BETA-ALANINE AND HISTIDINE AND IS FOUND IN ALMOST ALL VERTEBRATE MUSCLE, IS:
A. Carnosine
B. Creatine
C. Bradykinin
D. Ergothioneine
E. Carnitine
Ref. 1 - pp. 112, 582
Ref. 2 - p. 566

715. ESSENTIAL AMINO ACID(S) IN WHITE RATS:
A. L-proline
B. L-arginine
C. L-methionine
D. L-serine
E. All of these Ref. 2 - p. 349

SECTION IV - AMINO ACIDS AND PROTEINS

716. WHICH OF THE FOLLOWING IS NOT OPTICALLY ACTIVE?:
A. Leucine
B. Alanine
C. Glycine
D. Cysteine
E. Lysine
Ref. 1 - p. 94

717. A SULFUR-CONTAINING AMINO ACID IS:
A. Sulfinylpyruvic acid
B. Chondroitin sulfate
C. Homocysteine
D. Tryptophan
E. Glutathione
Ref. 1 - pp. 554, 584

718. WHICH OF THE FOLLOWING AMINO ACIDS POSSESSES AN IMINO GROUP?:
A. Tryptophan
B. Hydroxylysine
C. Tyrosine
D. Valine
E. Proline
Ref. 1 - p. 91

719. BETAINES ARE:
A. Phosphagens
B. Quaternary ammonium compounds
C. Purine-like compounds
D. Glycosides found in animal tissues
E. Lipid components of the central nervous system
Ref. 1 - p. 554
Ref. 3 - pp. 50, 807

720. AN AMINO ACID WHICH CONTAINS A DISULFIDE BOND IS:
A. Lysine
B. Methionine
C. Homocysteine
D. Cystine
E. Cysteine
Ref. 1 - p. 91

721. THE STRENGTH OF AN ACID IS MEASURED BY ITS:
A. Solubility product
B. Dissociation constant
C. Ability to donate electrons
D. Dielectric constant
E. Molarity
Ref. 1 - p. 96

722. THE ISOELECTRIC POINT OF AN AMINO ACID DEPENDS UPON ITS:
A. Optical rotation
B. Dissociation constant
C. Diffusion constant
D. Chain length
E. Birefringence
Ref. 1 - pp. 101, 105

723. IN THE PROCESS OF ELECTROPHORESIS, IF THE pH IS ABOVE THAT OF ITS ISOELECTRIC POINT, A PROTEIN WILL:
A. Migrate to the negative pole
B. Migrate to the positive pole
C. Not migrate
D. Form a zwitterion
E. Precipitate
Ref. 1 - pp. 118, 121-123

724. CHEMICALLY KERATIN IS A:
A. Globulin
B. Fibrous protein
C. Tripeptide
D. Histone
E. Conjugated protein
Ref. 1 - pp. 116-117

725. GLOBULINS AND ALBUMINS ARE DEFINED AS:
A. Derived proteins
B. Conjugated proteins
C. Globular proteins
D. Lipoproteins
E. Fibrous proteins
Ref. 1 - p. 117

726. A CERTAIN AMINO ACID, WHEN PRESENT IN A POLYPEPTIDE, WILL PRODUCE AN INTERRUPTION OF THE ALPHA-HELICAL STRUCTURE. THIS AMINO ACID IS:
A. Tyrosine
B. Tryptophan
C. Valine
D. Proline
E. Lysine
Ref. 5 - p. 54

727. THE ALPHA-HELICAL STRUCTURE OF CERTAIN PROTEINS:
A. Is due primarily to disulfide bridges
B. Is a function of complementary polypeptide chains
C. Is due primarily to hydrogen bonding
D. Is a property of all proteins
E. Is due to their DNA content
Ref. 5 - p. 54

728. THE TERM "FINGER-PRINTING" OF A PROTEIN IMPLIES:
A. Complete acid hydrolysis of the protein followed by gel electrophoresis
B. Tryptic digestion, followed by chromatography and electrophoresis
C. Paper chromatography and electrophoresis of the intact protein
D. Basic hydrolysis, followed by carboxypeptidase and agarose gel electrophoresis
E. Digestion with carboxypeptidase, followed by high voltage electrophoresis
Ref. 5 - p. 707

729. AN AMINO ACID NOT INVOLVED IN THE UREA SYNTHESIS CYCLE IS:
A. Arginine
B. Histidine
C. Citrulline
D. Ornithine
E. Aspartic acid
Ref. 1 - pp. 560-561

730. THE CONVERSION OF ALANINE TO CARBOHYDRATE IS TERMED:
A. Glycolysis
B. Oxidative decarboxylation
C. Specific dynamic action
D. Glycogenesis
E. Glycogenolysis
Ref. 1 - pp. 442, 595

731. ALL OF THE FOLLOWING ARE CATABOLIC PRODUCTS OF TRYPTOPHAN, EXCEPT:
A. Leucine
B. Xanthurenic acid
C. Nicotinic acid
D. Anthranilic acid
E. Kynurenine
Ref. 1 - p. 614

732. FORMATION OF MELANIN FROM TYROSINE REQUIRES THE ACTION OF:
A. Dopa decarboxylase
B. Diamine oxidase
C. Peroxidase
D. Tyrosinase
E. Catechol oxidase
Ref. 1 - p. 608

733. THE MAJOR CATEGORY OF PROTEINS INTO WHICH NEARLY ALL ANTIBODIES FALL IS:
A. Serum albumins
B. Beta-globulins
C. Lipoproteins
D. Gamma globulins
E. Glycoproteins
Ref. 1 - p. 710

SECTION IV - AMINO ACIDS AND PROTEINS

734. THE ALPHA-AMINO NITROGEN OF AMINO ACIDS APPEARS IN THE URINE OF MAMMALS AS:
A. Uric acid
B. Ammonia
C. Urea
D. Glutamine
E. Creatinine
Ref. 1 - pp. 560-562

735. REMOVAL OF AMINO GROUPS FROM AMINO ACIDS IN ANIMALS IS CARRIED OUT BY:
A. Hydrolysis and reductive amination
B. Transamination only
C. Oxidative deamination and transamination
D. Anaerobic deamination only
E. None of these
Ref. 1 - pp. 558-559

736. THE METABOLISM OF GLYCINE IS IMPORTANT IN THE FORMATION OF C_1 UNITS; A VITAMIN FUNCTIONING AS A CO-FACTOR IN GLYCINE METABOLISM IS:
A. Folic acid
B. Thiamine
C. Pantothenic acid
D. Vitamin E
E. Cobamide
Ref. 1 - p. 597

737. OXIDATIVE DEAMINATION OF AMINO ACIDS OCCURS IN TWO STEPS. AN INTERMEDIATE IN THE PROCESS IS:
A. Alpha-keto acid
B. Urea
C. Uric acid
D. Ammonia
E. Alpha-imino acid
Ref. 1 - pp. 559-560
Ref. 2 - p. 358

738. THE GLYCOGENIC ACTION OF SERINE MAY BE EXPLAINED BY:
A. Its stimulation of insulin production
B. Its conversion to pyruvic acid by the action of threonine aldolase
C. Its direct utilization in the Embden-Meyerhof pathway
D. Its decarboxylation to ethanolamine
E. None of these
Ref. 2 - pp. 371-372

739. IN THE NORMAL ANIMAL, THE SULFUR OF METHIONINE AND CYSTEINE IS EXCRETED MAINLY AS:
A. Conjugated esters
B. Inorganic sulfate
C. Organic sulfate
D. Sulfite
E. None of these
Ref. 1 - pp. 597, 603

740. IN URICOTELIC ANIMALS CARBAMYL GROUPS ARE TRANSFERRED TO ASPARATATE AND THEN UTILIZED IN PURINE SYNTHESIS. IN UREOTELIC ANIMALS CARBAMYL GROUPS ARE TRANSFERRED TO ORNITHINE TO FORM:
A. Pyrimidines
B. Urea
C. Citrulline
D. Uric acid
E. Arginine
Ref. 1 - pp. 545, 562

741. ADRENALINE AND NOR-ADRENALINE ARE FORMED FROM THE AMINO ACID:
A. Lysine
B. Tyrosine
C. Proline
D. Iso-leucine
E. Tryptophan
Ref. 5 - pp. 440-442

742. THE REACTION WITH NINHYDRIN (FORMATION OF PURPLE COLOR) IS CHARACTERISTIC OF WHICH OF THE FOLLOWING?:
A. Peptides and proteins, but not free amino acids
B. Most primary amines
C. Alpha-amino acids, peptides and proteins
D. Small peptides and alpha-amino acids, but not proteins
E. Hydroxyproline
Ref. 2 - p. 69

743. THE ISO-ELECTRIC POINT OF ALANINE IS 6.0. IF ALANINE IS DISSOLVED IN A BUFFER OF pH 3.0 AND SUBJECTED TO ELECTROPHORESIS:
A. It will not migrate to either anode or cathode
B. It will migrate to the cathode
C. It will decompose with evolution of hydrogen
D. Some will migrate to the anode and some to the cathode
E. It will migrate to the anode
Ref. 2 - p. 65

744. WHICH OF THE FOLLOWING STATEMENTS CONCERNING D-AMINO ACIDS IS CORRECT?:
A. D-amino acids cannot be metabolized by mammals
B. D-amino acids are not found in plants or microorganisms
C. D-amino acids can be transformed into L-amino acids in the mammalian organism
D. D-amino acids can be transformed directly into the L-form without involving a ketonic intermediate
E. D-amino acids cannot be utilized for the formation of the bacterial cell wall
Ref. 5 - pp. 403-404

745. AN ENZYME IS PRESENT IN BRAIN WHICH CONVERTS GLUTAMIC ACID TO GAMMA-AMINOBUTYRIC ACID. THIS REACTION REQUIRES:
A. Pyridoxal phosphate
B. ATP
C. Serine
D. Carbon dioxide
E. Glutamine
Ref. 1 - pp. 585-586, 862

746. THE METABOLISM OF L-TRYPTOPHAN RESULTS IN A SUBSTANCE WHICH MAY CAUSE:
A. A fall in blood pressure
B. Sweating
C. Vasoconstriction
D. Albinism
E. Phenylketonuria
Ref. 1 - p. 589

747. THE MAJOR SOURCE OF AMMONIA PRODUCED BY THE KIDNEYS IS:
A. Leucine
B. Glycine
C. Alanine
D. Glutamine
E. Asparagine
Ref. 1 - pp. 596, 795

748. DECARBOXYLATION OF WHICH OF THE FOLLOWING AMINO ACIDS RESULTS IN FORMATION OF A VASODILATOR?:
A. L-glutamic acid
B. L-arginine
C. L-histidine
D. L-aspartic acid
E. L-valine
Ref. 1 - p. 588

SECTION IV - AMINO ACIDS AND PROTEINS

749. ESSENTIAL AMINO ACIDS ARE:
A. Not synthesized by the organism at a rate adequate to meet metabolic requirements and must be supplied in the diet
B. Synthesized in the kidney but cannot be formed in other tissues
C. Not excreted by man because they are too important to the organism
D. The same for growing and adult animals
E. Metabolized by D-amino acid oxidase
Ref. 1 - p. 540

750. THYROXINE IS DERIVED FROM:
A. Tyrosine
B. Tyramine
C. Taurine
D. Tryptamine
E. Tryptophan
Ref. 1 - pp. 922 -923

751. TRANSFER OF METHYL GROUPS FROM CHOLINE TO HOMOCYSTEINE RESULTS IN THE FORMATION OF:
A. Cystine
B. Ethanolamine
C. Betaine
D. Methionine + ethanolamine
E. Cysteic acid
Ref. 2 - pp. 352-354

752. AN ANIMAL IS IN NEGATIVE NITROGEN BALANCE WHEN:
A. Intake exceeds output
B. New tissue is being synthesized
C. Output exceeds intake
D. Intake is equal to output
E. The urine is nitrogen-free
Ref. 1 - p. 555

753. THE ONLY KNOWN PHYSIOLOGICAL METHYLATING AGENTS IN THE ANIMAL ORGANISM ARE:
A. Choline and betaine
B. Choline and S-adenosylmethionine
C. Betaine and S-adenosylmethionine
D. Dimethylglycine and betaine
E. Betaine aldehyde and methionine
Ref. 1 - pp. 583-584

754. A MAJOR METABOLIC PATHWAY OF TRYPTOPHAN INITIALLY INVOLVES THE FORMATION OF:
A. Nicotinic acid
B. Hydroxyindoleacetic acid
C. Serotonin
D. Kynurenine
E. Anthranilic acid
Ref. 1 - p. 614

755. TWO VITAMINS INVOLVED IN THE TRANSFORMATION OF SERINE TO GLYCINE:
A. B_{12} + nicotinamide
B. Folic acid + B_6
C. Pyridoxal phosphate + B_{12}
D. Folic acid + pyridoxal phosphate
E. B_{12} + folic acid
Ref. 1 - pp. 550-551

756. PHENYLKETONURIA, AN INBORN ERROR OF METABOLISM, IS DUE TO THE ABSENCE OF THE ENZYME WHICH:
A. Deaminates tyrosine
B. Hydroxylates phenylalanine
C. Oxidizes homogentisic acid
D. Converts tyrosine to dopa
E. Iodinates tyrosine
Ref. 1 - p. 549

757. ALKAPTONURIA, AN INBORN ERROR OF METABOLISM, IS DUE TO THE ABSENCE OF THE ENZYME:

A. Tyrosinase
B. Phenylalanine hydroxylase
C. Dopa decarboxylase
D. Homogentisic acid oxidase
E. Dopa oxidase

Ref. 1 - p. 608

758. THE HELICAL CONTENT OF POLYPEPTIDES:

A. Can be estimated by analyzing for pleated sheet structures
B. Can be estimated by optical rotatory dispersion
C. Can be estimated by ion exchange chromatography
D. Is minimal in the case of lactoglobulin
E. Cannot be measured with methods currently available

Ref. 2 - p. 79

759. AN IMPORTANT STRUCTURAL FEATURE OF ANTIBODIES:

A. They are composed of 2 peptide chains
B. They are composed of a single peptide
C. They are composed of 2 light and 2 heavy chains
D. They cannot be dissociated by urea
E. They contain no disulfide bridges

Ref. 3 - p. 129

760. APOLAR BONDS:

A. Are of little importance in protein structure
B. Are neutralized polar bonds
C. Cannot form in aqueous solution
D. Represent an interaction between hydrocarbons and peptide bonds
E. Are also known as hydrophobic bonds

Ref. 3 - p. 165

761. ISOZYMES:

A. Are enzymes that exist in more than one structural form in the same species
B. Cannot be distinguished in a given species except immunologically
C. By definition must have identical amino acid composition
D. Are single polypeptide chains which differ in composition by a single amino acid
E. Have identical catalytic properties

Ref. 3 - p. 174

762. ALBINISM, AN INBORN ERROR OF METABOLISM, IS DUE TO THE LACK OF:

A. Tyrosinase
B. Homogentisicase
C. Kynureninase
D. Phenylalanine hydroxylase
E. Tyrosine decarboxylase

Ref. 1 - p. 608

763. WHICH OF THE FOLLOWING AMINO ACIDS UNDERGOES TRANSAMINATION TO FORM OXALOACETIC ACID?:

A. Alanine
B. Aspartic acid
C. Glutamic acid
D. Proline
E. Threonine

Ref. 2 - p. 351

764. JACOB AND MONOD POSTULATED THAT ENZYME "INDUCERS" EXERT THEIR EFFECT BY REACTING WITH:
A. Aminoacyl-tRNA
B. The regulatory metabolite
C. A structural gene
D. A repressor protein
E. An operator gene
Ref. 2 - p. 105

765. ACTINOMYCIN D INTERFERES WITH ENZYME INDUCTION BY:
A. Combining with the repressor protein
B. Combining with ribosomal RNA
C. Combining with DNA, thus preventing formation of mRNA
D. Combining with the negative feedback system
E. Combining with tRNA
Ref. 5 - pp. 46-47

766. WHICH ENZYME WOULD SPLIT THE PEPTIDE LEUCYLGLYCYLGLYCINE INTO LEUCINE AND GLYCYLGLYCINE?:
A. Carboxypeptidase
B. Glycylglycyldipeptidase
C. Aminopeptidase
D. Chymotrypsin
E. Trypsin
Ref. 2 - pp. 340-344

767. WHICH AMINO ACID POSSESSES NO ASYMMETRIC CARBON ATOM?:
A. Glycine
B. Serine
C. Leucine
D. Valine
E. Histidine
Ref. 2 - p. 63

768. THE MAJOR END PRODUCT OF PROTEIN-NITROGEN METABOLISM IN MAN IS:
A. Glycine
B. Uric acid
C. Urea
D. Ammonia
E. Allantoin
Ref. 2 - pp. 361-362

769. NINHYDRIN IS A REAGENT WHICH GIVES A COLOR:
A. At the isoelectric point
B. With amino acids, peptides and proteins
C. With peptide bonds only
D. With secondary amino acids only
E. With polysaccharides and proteins
Ref. 1 - pp. 106, 136

770. THE TERMS "PRIMARY," "SECONDARY" AND "TERTIARY" STRUCTURE OF A PROTEIN REFER TO ITS:
A. Alpha-helix, beta-helix and gamma-helix
B. Peptide bonds and disulfide bonds
C. Peptide bonds, hydrogen bonds and non-covalent bonds
D. Electrostatic interaction, van der Waals forces and disulfide bonds
E. Amino acid sequence, hydrogen bonds and peptide bonds
Ref. 1 - pp. 138-139
Ref. 4 - pp. 38-40

771. IN THE DETERMINATION OF PROTEIN STRUCTURE PERFORMIC ACID OXIDATION IS USED TO:
A. Obtain information on conformation
B. Break the disulfide bonds
C. Denature the protein
D. Determine the C-terminal amino acid
E. Hydrolyze the protein
Ref. 1 - pp. 151, 147
Ref. 4 - pp. 36-37

ANSWER THE FOLLOWING QUESTIONS BY USING THE KEY OUTLINED BELOW:
A. If A is greater than B
B. If B is greater than A
C. If A and B are equal or nearly equal

772. A. Cation concentration in solution of an amino acid at its isoelectric point
B. Anion concentration in solution of an amino acid at its isoelectric point
Ref. 3 - pp. 107-108
Ref. 4 - pp. 26-27

773. A. Buffering capacity of an amino acid solution at isoelectric point
B. Buffering capacity of an amino acid solution at its pK
Ref. 1 - p. 100

774. A. pH at isoelectric point of lysine
B. pH at isoelectric point of glycine
Ref. 1 - pp. 101-103

775. A. Solubility of globulins in aqueous solutions
B. Solubility of globulins in aqueous solutions containing neutral salts
Ref. 1 - p. 124

776. A. pK value of alpha-amino groups in proteins
B. pK value of epsilon-amino groups in proteins
Ref. 1 - p. 120

777. A. Hexosamine content of mucoproteins
B. Hexosamine content of glycoproteins
Ref. 1 - pp. 117-118

778. A. Solubility of albumins in water
B. Solubility of globulins in water
Ref. 1 - p. 117

779. A. Solubility of collagen in boiling alkali
B. Solubility of collagen in boiling water
Ref. 1 - p. 116

780. A. Hydroxyproline content of collagen
B. Hydroxyproline content of protamines
Ref. 1 - pp. 116-117

781. A. Basicity of albumins
B. Basicity of protamines
Ref. 1 - p. 117

782. A. Carbohydrate content of mucoproteins
B. Carbohydrate content of glycoproteins
Ref. 1 - pp. 117-118

783. A. Sialic acid content of mucoproteins
B. Sialic acid content of lipoproteins
Ref. 1 - p. 117-118

784. A. pK of gamma-carboxyl groups in proteins
B. pK of phenolic hydroxyl groups in proteins
Ref. 1 - p. 120

785. A. Solubility of lipoproteins in water
B. Solubility of proteolipids in water Ref. 1 - p. 118

786. A. Number of potential asymmetric centers in hydroxylysine
B. Number of potential asymmetric centers in threonine
Ref. 1 - p. 95

787. A. Formation of trichloracetic acid salts of proteins at pH above isoelectric point
B. Binding of heavy metals by proteins at pH above isoelectric point
Ref. 1 - pp. 120-121

ANSWER THE FOLLOWING QUESTIONS BY USING THE KEY OUTLINED BELOW:
A. If 1, 2 and 3 are correct
B. If 1 and 3 are correct
C. If 2 and 4 are correct
D. If only 4 is correct
E. If all four are correct

788.
1. Proteins exhibit amphoteric properties
2. Proteins are not affected by pH changes, heat or ultraviolet radiation
3. Proteins contain free amino and carboxyl groups derived, for example, from gamma-carboxyl groups of glutamic acid and epsilon-amino groups of lysine
4. The specific properties of proteins are due almost entirely to the peptide bonds Ref. 1 - pp. 87-88; 118-119

789. PROTEINS:
1. Contain carbon, hydrogen, oxygen and nitrogen
2. Contain approximately 16% nitrogen by weight
3. Can be hydrolyzed to ammonia and free amino acids
4. Are composed primarily of alpha-amino acids
Ref. 1 - pp. 119-121

790.
1. The biuret reaction is given by all alpha-amino acids
2. Glutathione contains glutamic acid, glycine and histidine
3. Dipeptides usually give a positive biuret test
4. Tripeptides and proteins usually give a positive biuret test
Ref. 1 - pp. 111-112

791.
1. The major linkage between amino acids in proteins is the hydrogen bond
2. Intact proteins show little free amino nitrogen, but large amounts are formed by hydrolysis
3. Acid hydrolysis of proteins yields very unequal amounts of amino and carboxyl groups
4. Partial hydrolysis of proteins yields amino acids and polypeptides
Ref. 1 - pp. 110-111

792. 1. The Millon reaction is given by tyrosine-containing proteins
2. Formaldehyde combines with uncharged amino groups to form weakly basic derivatives
3. Cysteine and proteins with free sulfhydryl groups give a red color with sodium nitroprusside in dilute NH_3
4. Pure proteins possess a sharp melting point

Ref. 1 - pp. 102, 107-107, 161

EACH GROUP OF QUESTIONS BELOW CONSISTS OF FIVE LETTERED HEADINGS FOLLOWED BY A LIST OF NUMBERED WORDS OR PHRASES. FOR EACH NUMBERED WORD OR PHRASE SELECT THE ONE LETTERED HEADING THAT IS MOST CLOSELY RELATED TO IT:

A. Uropepsin
B. Pepsin
C. Rennin
D. Trypsin
E. α-Chymotrypsin

793. ___ The stable, active form of chymotrypsin
794. ___ Urinary pepsinogen
795. ___ Casein → paracasein
796. ___ Proteinase with pH optimum 1.5-2.2
797. ___ Zymogen converted to active enzyme by enterokinase

Ref. 2 - pp. 339-342

A. Beta-alanine
B. Ornithine
C. Dihydroxyphenylalanine
D. Hydroxylysine
E. 3, 5, 3'-triiodothyronine

798. ___ Dopa
799. ___ Found in the important vitamin pantothenic acid
800. ___ Involved in the formation of arginine
801. ___ Has been found only in collagen and gelatin
802. ___ Has hormonal activity

Ref. 1 - pp. 93-94, 545, 953

A. 1-fluoro-2, 4-dinitrobenzene (FDNB)
B. Phosgene
C. Phenylisothiocyanate
D. 1, 2-naphthoquinone-4-sulfonate
E. Sulfonated polystyrene resin

803. ___ Used in Bergmann-Zervas method of peptide synthesis
804. ___ Used in automatic amino acid analyzers
805. ___ Used in end-group analysis
806. ___ Used in Folin's reaction for amino acids
807. ___ Edman's reagent

Ref. 1 - pp. 107-109, 113-115, 142-143

MATCH THE FOLLOWING AMINO ACIDS WITH THEIR CORRECT CONVERSION PRODUCTS:

A. Tryptophan
B. Cysteine
C. Tyrosine
D. Serine
E. Proline

808. ___ L-Glutamate
809. ___ Adrenalin
810. ___ Taurine
811. ___ Serotonin
812. ___ Ethanolamine

Ref. 2 - p. 381
Ref. 5 - pp. 394, 426, 440, 445

THE REAGENTS LISTED BELOW GIVE A RED COLOR WITH SPECIFIC AMINO ACIDS. MATCH THE REACTION WITH THE AMINO ACID IT DETECTS:

A. Sodium nitroprusside in dilute NH_4OH
B. Mercurous nitrate in nitric acid
C. Diazotized sulfanilic acid in alkaline medium
D. α-Naphthol and NaOCl
E. Glyoxylic acid in conc. H_2SO_4

813. ___ Tyrosine
814. ___ Cysteine
815. ___ Arginine
816. ___ Histidine
817. ___ Tryptophan

Ref. 2 - p. 69

A. Phenylisothiocyanate method
B. Dinitrofluorobenzene method
C. Urea and guanidine
D. Cystine
E. Hemoglobin

818. ___ Used to identify terminal amino acids in proteins without hydrolysis of remainder of the peptide chain
819. ___ Used to identify free amino groups in proteins, requires complete hydrolysis of the protein
820. ___ Contain(s) four polypeptide chains
821. ___ Disrupt(s) hydrogen bonds of proteins thus causing drastic alterations in conformation
822. ___ Link(s) the two peptide chains of insulin

Ref. 1 - pp. 142-144, 151, 161-163

A. Hydrogen bond
B. Covalent bond
C. Alpha-helix
D. Denaturation
E. Hydrophobic bond

823. ___ Formed between R groups of certain amino acids such as alanine, valine, phenylalanine, tyrosine
824. ___ Associated with increased levorotation
825. ___ Existence verified by X-ray diffraction
826. ___ Peptide bond
827. ___ Due to tendency of a hydrogen atom to share electrons of an oxygen atom

Ref. 1 - pp. 107, 152-155, 161

MATCH THE FORMULAS WITH THE CORRECT NAME:

A.

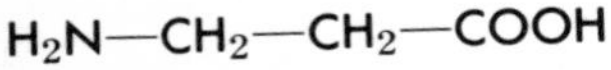

$$H_2N—CH_2—CH_2—COOH$$

B.

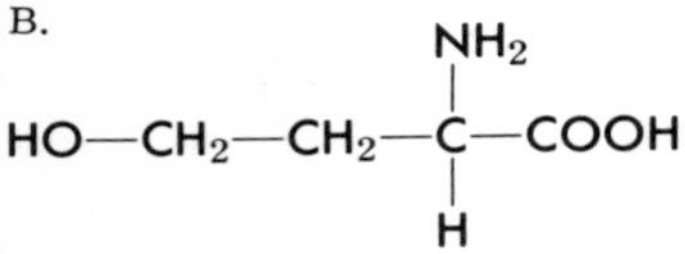

$$HO—CH_2—CH_2—\overset{\displaystyle NH_2}{\underset{\displaystyle H}{C}}—COOH$$

C.

$$H_2N—CH_2—CH_2—CH_2—COOH$$

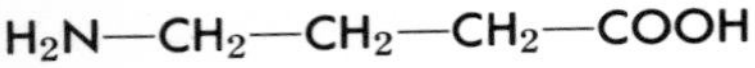

D.

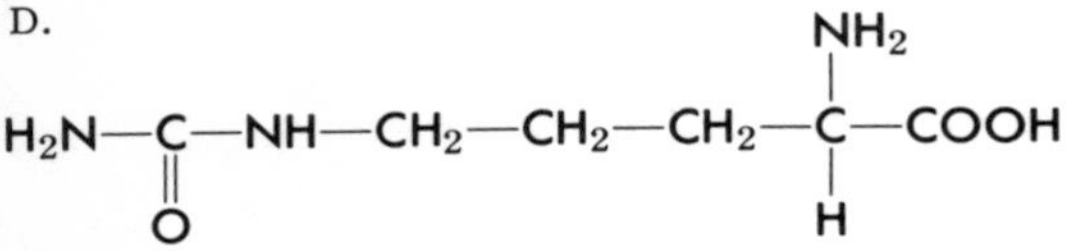

$$H_2N—\underset{\displaystyle O}{\overset{}{\underset{\|}{C}}}—NH—CH_2—CH_2—CH_2—\overset{\displaystyle NH_2}{\underset{\displaystyle H}{C}}—COOH$$

E.

$$H_2N—CH_2—CH_2—CH_2—\overset{\displaystyle NH_2}{\underset{\displaystyle H}{C}}—COOH$$

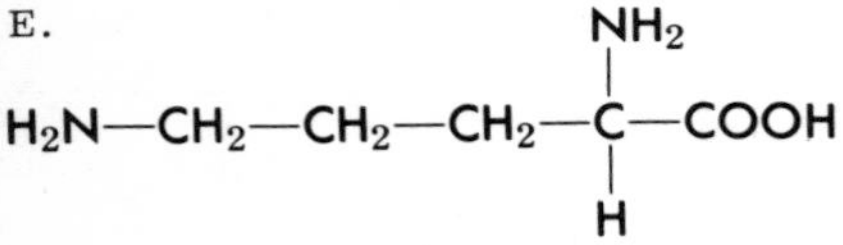

828. ___ Ornithine

829. ___ Citrulline

830. ___ Gamma-Aminobutyric acid

831. ___ Homoserine

832. ___ β-Alanine

Ref. 1 - pp. 93; 150; 545; 586

MATCH THE FORMULA WITH THE CORRECT NAME:

A.

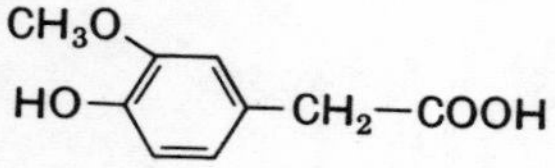

CH_3O
HO— (ring) —CH_2-COOH

B.

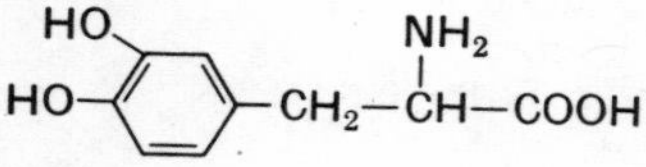

HO NH_2
HO— (ring) —$CH_2-CH-COOH$

C.

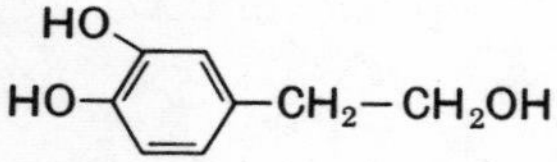

HO
HO— (ring) —CH_2-CH_2OH

D.

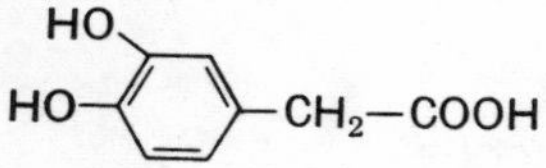

HO
HO— (ring) —CH_2-COOH

E.

HO
HO— (ring) —CH_2-CH_2
NH_2

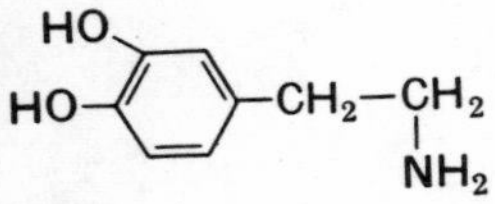

833. ___ Homovanillic acid

834. ___ Dopa

835. ___ 3, 4-Dihydroxyphenylethanol

836. ___ Hydroxytyramine

837. ___ Homoprotocatechuic acid

Ref. 1 - p. 589

MATCH THE AMINO ACIDS LISTED BELOW WITH THEIR CORRECT BIOLOGICAL PRECURSORS:

A. Pyruvate
B. Phospho-glycerate
C. Erythrose-4-phosphate
D. Oxalacetate
E. Meso-diaminopimelic acid

838. ___ Aspartic acid
839. ___ Tyrosine
840. ___ Lysine
841. ___ Serine
842. ___ Alanine

Ref. 3 - p. 792

MATCH THE FORMULAS WITH THE CORRECT NAME:

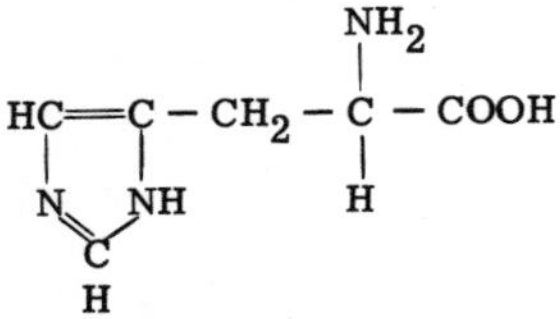

A. $HC{=}C(-N{=}CH-NH-)-CH_2-CH(NH_2)-COOH$ (imidazole ring: HC, C, N, NH, C–H)

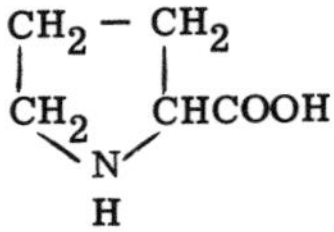

B. CH_2-CH_2, CH_2 $CHCOOH$, N–H (pyrrolidine ring)

C. $C_6H_5-CH_2-CH(NH_2)-COOH$

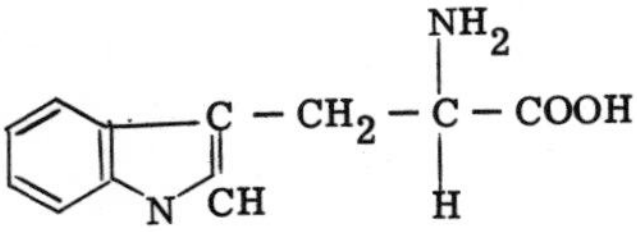

D. (indole ring) $C-CH_2-CH(NH_2)-COOH$, with N and CH in the ring

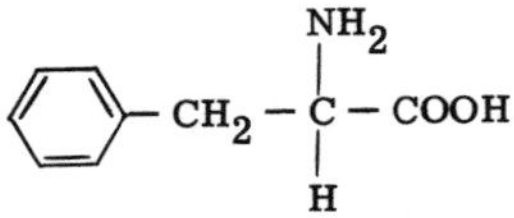

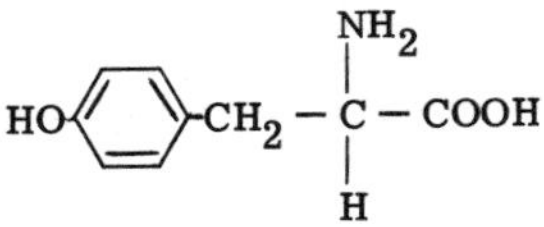

E. $HO-C_6H_4-CH_2-CH(NH_2)-COOH$

843. ___ Proline

844. ___ Phenylalanine

845. ___ Tyrosine

846. ___ Tryptophan

847. ___ Histidine

Ref. 1 - pp. 89-93

SECTION IV - AMINO ACIDS AND PROTEINS

ANSWER THE FOLLOWING QUESTIONS BY USING THE KEY OUTLINED BELOW:
A. If the item is associated with A only
B. If the item is associated with B only
C. If the item is associated with both A and B
D. If the item is associated with neither A nor B

IN THE CASE OF PROTEINS:

A. Solubility
B. Osmotic pressure
C. Both
D. Neither

848. ___ Is dependent upon the molecular weight
849. ___ Is dependent upon pH
850. ___ Is influenced by ionic strength of solution
851. ___ Is at a minimum at the isoelectric point
852. ___ Can be utilized to determine purity of a protein

Ref. 1 - pp. 123; 127-128
Ref. 4 - pp. 5-10

ANSWER THE FOLLOWING QUESTIONS BY USING THE KEY OUTLINED BELOW:
A. If 1, 2 and 3 are correct
B. If 1 and 3 are correct
C. If 2 and 4 are correct
D. If only 4 is correct
E. If all four are correct

853. THE ESSENTIAL AMINO ACIDS:
1. Must be supplied in the diet because the organism has lost the capacity to aminate the corresponding keto-acids
2. Must be supplied in the diet because the organism has an impaired ability to synthesize the carbon chain of the corresponding keto-acids
3. Are identical in all mammals studied
4. Are defined as "those amino acids which cannot be synthesized by the organism at a rate adequate to meet metabolic requirements"

Ref. 1 - pp. 540; 1003

854. THE ENZYME PEPSIN:
1. Participates in an autocatalytic process since in the presence of acid it catalyzes the conversion of pepsinogen to pepsin
2. Acts primarily on peptide bonds involving aromatic amino acids
3. Is less active in pernicious anemia so that peptic digestion in the stomach is limited, because the pH is above the pH optimum of pepsin
4. Although present, in achylia gastrica is unable to exert its proteolytic effect because acid is absent from the gastric contents

Ref. 1 - p. 532

855. DURING INTESTINAL PROTEOLYSIS:
1. The enzyme enterokinase converts trypsinogen to trypsin
2. The activation of trypsinogen to trypsin involves no change in enzyme structure or conformation
3. Trypsin is an activator of procarboxypeptidases and proelastase
4. Unless pepsin has first acted upon dietary protein, trypsin and chymotrypsin alone cannot produce free amino acids from the protein

Ref. 1 - pp. 533-534

856. AFTER INTESTINAL PROTEOLYSIS IS COMPLETED:
1. The amino acids produced leave the intestine, entering lymph and portal blood to about equal extent
2. Amino acids, peptides and even native protein can be absorbed from the intestine
3. Contrary to expectation, D- and L-amino acids are absorbed at about equal rates
4. Studies in the rat have shown that vitamin B_6 influences intestinal absorption of amino acids

Ref. 1 - pp. 534-535

857. NITROGEN IS CONVERTED INTO ORGANIC COMPOUNDS BY BOTH PLANTS AND MAMMALS:
1. Nitrate reductase, a molybdoflavoprotein, is present in mammalian cells
2. NH_3 is fixed at all phylogenetic levels by 3 major reactions, namely the syntheses of glutamic acid, glutamine and urea
3. Glutamic acid synthesis from alpha-ketoglutarate requires the presence of ATP
4. Carbamyl phosphate synthesis in mammalian liver has an absolute requirement for N-acetylglutamic acid

Ref. 1 - pp. 535-537

858. REGARDING TRANSMETHYLATION:
1. Methionine is the methyl donor only indirectly since it must first be transformed into the active form, S-adenosylmethionine
2. During the activation process ATP is required and the phosphate radicals of ATP are converted to P_i and PP_i
3. During the methyl transfer methionine is converted to homocysteine
4. Mammalian tissues can regenerate methionine, a process involving lecithin

Ref. 1 - pp. 553-554

859. IN THE BIOSYNTHESIS OF ESSENTIAL AMINO ACIDS:
1. The formation of methionine in plants and microorganisms involves the reduction of $SO_4^{=}$
2. The carbon atoms of phenylalanine are derived from erythrose-4-phosphate and phosphoenolpyruvate
3. 5-phosphoshikimic acid is an intermediate in phenylalanine and tryptophan biosynthesis
4. Serine serves as the precursor of threonine and lysine

Ref. 1 - pp. 563; 573-575

860. AMINO ACIDS SERVE NOT ONLY AS PRECURSORS OF PROTEINS BUT ALSO OF OTHER COMPOUNDS. IMPORTANT EXAMPLES OF THE LATTER PATHWAYS:
1. The formation of carnosine from histidine and beta-alanine
2. The role of alpha-aminobutyric acid in porphyrin biosynthesis
3. The formation of serotonin from tryptophan
4. The formation of alpha-aminobutyric acid from threonine

Ref. 1 - pp. 579; 589

861. PHENYLKETONURIA:
1. Is associated with the development of mental retardation. This can be prevented by feeding tyrosine
2. Is caused by a hereditary lack of phenylalanine hydroxylase
3. Is caused by a genetic defect associated with an overproduction of tyrosine dehydroxylase
4. Phenylalanine hydroxylase, a microsomal enzyme, requires TPNH to ensure a supply of a reduced pteridine as cofactor

Ref. 1 - pp. 548-549; 606; 953

EACH GROUP OF QUESTIONS BELOW CONSISTS OF 5 LETTERED HEADINGS FOLLOWED BY A LIST OF NUMBERED WORDS OR PHRASES. FOR EACH NUMBERED WORD OR PHRASE SELECT THE ONE LETTERED HEADING THAT IS MOST CLOSELY RELATED TO IT:

A. $NH_2\overset{NH}{\overset{||}{C}}-\overset{CH_3}{\overset{|}{N}}-CH_2COOH$

B. $NH_2\overset{NH}{\overset{||}{C}}-NH-CH_2CH_2CH_2CHNH_2COOH$

C. $NH_2\overset{O}{\overset{||}{C}}-CH_2CH_2CHNH_2COOH$

D. $NH_2CH_2CH_2CH_2CHNH_2COOH$

E. $NH_2CH_2CH_2CH_2CH_2CHNH_2COOH$

862.___ Action of arginase on _______ produces urea
863.___ Essential amino acid
864.___ Major precursor of urinary ammonia
865.___ Biosynthesis from glycocyamine
866.___ Formed from arginine in urea synthesis cycle

Ref. 1 - pp. 540; 560-561; 581-582; 596

A. $HS-CH_2-CH(NH_2)COOH$
B. $HS-CH_2-CH_2CH(NH_2)COOH$
C. $CH_3-S-CH_2CH_2CH(NH_2)COOH$
D. $(SO_3H)CH_2CH_2NH_2$
E. $HO_2S-CH_2-CH(NH_2)COOH$

867. ___ Constituent of glutathione
868. ___ Demethylation product of methionine
869. ___ Produced by action of cystathionase
870. ___ Produced by decarboxylation of cysteic acid
871. ___ Constituent of bile salts
872. ___ Precursor of cysteic acid
873. ___ Precursor of beta-sulfinylpyruvic acid

All in Ref. 1
867 - pp. 578-579
868 - p. 603
869 - p. 547
870 - p. 600
871 - p. 600
872 - p. 600
873 - p. 600

FIND THE CORRECT ENZYMATIC DECARBOXYLATION PRODUCTS OF THE AMINO ACIDS LISTED BELOW:

A. Agmatine
B. β-Alanine
C. Tyramine
D. Isobutylamine
E. Isoamylamine

874. ___ L-aspartic acid
875. ___ L-valine
876. ___ L-Leucine
877. ___ L-arginine
878. ___ L-tyrosine

WHAT ARE THE PRODUCTS OF THE FOLLOWING ENZYMATIC REACTIONS?:

A. Glycine
B. Creatine
C. Cystathionine
D. Taurine
E. Isethionic acid

879. ___ Guanidoacetic acid + S-adenosylmethonine
880. ___ Taurine $\rightarrow$
881. ___ Cysteic acid $\rightarrow$ CO_2 + __________
882. ___ Homocysteine + pyridoxal phosphate + serine + transsulfurase $\rightarrow$
883. ___ Serine + Tetrahydrofolic acid + pyridoxal phosphate + Mn^{++} $\rightarrow$

All in Ref. 1
879 - p. 581 882 - p. 547
880 - p. 601 883 - p. 551
881 - p. 600

CLASSIFY THE FOLLOWING AMINO ACIDS ACCORDING TO THEIR GLYCOGENIC AND/OR KETOGENIC EFFECTS:

A. Ketogenic
B. Glycogenic
C. Ketogenic and glycogenic

884. ___ Alanine
885. ___ Leucine
886. ___ Tyrosine
887. ___ Glutamic acid
888. ___ Isoleucine

Ref. 1 - p. 595

MATCH THE NUMBERED COMPOUNDS (QUESTIONS 889-893) WITH THE MOST CORRECT PRECURSOR:

A. Tyrosine
B. Shikimic acid
C. Lysine
D. N'-(5'-phosphoribosyl) ATP
E. Serine

889. ___ Histidine
890. ___ Homogentisic acid
891. ___ 3- Phosphoglyceric acid
892. ___ Diaminopimelic acid
893. ___ Phenylalanine

All in Ref. 1
889 - p. 576
890 - p. 574
891 - p. 549
892 - p. 569
893 - p. 607

MATCH THE FOLLOWING AMINO ACIDS WITH THEIR PROPER METABOLIC PRODUCTS:

A. Nicotinic acid
B. Gamma-aminobutyric acid
C. Homogentisic acid
D. Alpha-ketobutyric acid
E. Beta-hydroxy-beta-methylglutaryl-CoA

894. ___ Threonine
895. ___ Leucine
896. ___ Tyrosine
897. ___ Glutamic acid
898. ___ Tryptophan

All in Ref. 1
894 - p. 603
895 - p. 605
896 - p. 607
897 - p. 586
898 - p. 615

BIOSYNTHESIS OF THE FOLLOWING AMINO ACIDS REQUIRES WHICH FACTORS ?:

A. Anthranilic acid
B. Imidazoleglycerol phosphate
C. Acetolactic acid
D. Erythrose-4-phosphate
E. Aspartic acid semialdehyde

899. ___ Histidine
900. ___ Lysine
901. ___ Valine
902. ___ Tryptophan
903. ___ Phenylalanine

All in Ref. 1
899 - p. 576
900 - p. 569
901 - p. 571
902 - p. 575
903 - p. 574

FILL IN THE BLANKS APPROPRIATELY:

A. SCN^-
B. Inorganic $SO_4^=$
C. 3'-Phosphoadenosine-5'-phosphosulfate
D. $S_2O_3^=$
E. Isovalthine

904. ___ Formation of phenolic sulfates which are excreted in the urine requires ______________
905. ___ In normal animals the sulfur of methionine and cysteine is excreted primarily as ______________
906. ___ A reaction between inorganic sulfite and beta-mercaptopyruvate forms ______________ which is excreted in the urine
907. ___ The enzyme rhodanese catalyzes a reaction between thiosulfate and cyanide to form ______________
908. ___ ______________ was isolated from urine of hypercholesterolemic subjects

Ref. 1 - pp. 563; 597-598; 601-602

FIND THE CORRECT ENZYMATIC DECARBOXYLATION PRODUCTS OF THE AMINO ACIDS LISTED BELOW:

A. Cadaverine
B. Putrescine
C. Taurine
D. γ-Aminobutyric acid
E. Histamine

909. ___ L-lysine
910. ___ L-cysteic acid
911. ___ L-histidine
912. ___ L-ornithine
913. ___ L-glutamic acid

Ref. 3 - p. 794

ANSWER THE FOLLOWING QUESTIONS BY USING THE KEY OUTLINED BELOW:
A. If A is greater than B
B. If B is greater than A
C. If A and B are equal or nearly equal

914. RATE OF SYNTHESIS BY GROWING ALBINO RAT OF:
A. Lysine
B. Alanine
Ref. 1 - p. 540

915. NUTRITIONAL REQUIREMENT IN ALBINO RAT FOR:
A. Aspartic acid
B. Glutamic acid
Ref. 1 - p. 540

916. A. Storage capacity for fatty acids in mammals
B. Storage capacity for amino acids in mammals
Ref. 1 - p. 555

917. A. Uric acid excretion of man
B. Urea excretion of man
Ref. 2 - p. 727

918. A. Biosynthesis of serum proteins in liver
B. Biosynthesis of serum proteins in kidney
Ref. 1 - p. 722

919. A. Xanthurenic acid excretion in pyridoxine deficiency
B. Xanthurenic acid excretion in normal subject
Ref. 1 - p. 615

920. A. Biological half-life of collagen
B. Biological half-life of serum albumin
Ref. 1 - pp. 557-558

921. A. Decarboxylation of N-acetyl-L-aspartate in brain
B. Decarboxylation of alpha-aminobutyric acid in brain
Ref. 1 - p. 862

922. A. Number of carbon atoms per molecule of histamine
B. Number of carbon atoms per molecule of histidine
Ref. 1 - p. 613

923. A. Urea nitrogen excretion on a low protein diet
B. Urea nitrogen excretion on a high protein diet
Ref. 1 - pp. 555; 560

SECTION V - NUCLEOTIDES, PURINES, PYRIMIDINES

EACH OF THE QUESTIONS OR INCOMPLETE STATEMENTS BELOW IS FOLLOWED BY 5 SUGGESTED ANSWERS OR COMPLETIONS. SELECT THE ONE WHICH IS BEST IN EACH CASE:

924. PYRIMIDINES ARE OF BIOLOGICAL IMPORTANCE IN AREAS OTHER THAN NUCLEIC ACIDS. ONE SUCH PYRIMIDINE DERIVATIVE IS:
A. Adenine
B. Uric acid
C. Inosinic acid
D. Thiamine
E. Diphosphopyridine nucleotide Ref. 1 - p. 182

925. WHICH OF THE FOLLOWING STATEMENTS ABOUT NUCLEIC ACIDS IS MOST CORRECT?:
A. Both pentose nucleic acids and deoxypentose nucleic acids contain the same purines
B. Both pentose nucleic acids and deoxypentose nucleic acids contain the same pyrimidines
C. Cyclic AMP has been found in RNA
D. RNA contains cytosine and thymine
E. DNA and RNA are hydrolyzed by weak alkali
Ref. 1 - pp. 190-191; 201-202

926. ACID HYDROLYSIS OF RIBONUCLEIC ACID WOULD YIELD THE FOLLOWING MAJOR PRODUCTS:
A. D-deoxyribose, cytosine, adenine
B. D-ribose, thymine, guanine
C. D-ribose, cytosine, uracil, thymine
D. D-ribose, uracil, adenine, guanine, cytosine
E. D-deoxyribose, cytosine, thymine, adenine, guanine
Ref. 1 - p. 202

927. COMPLETE ACID HYDROLYSIS OF NUCLEIC ACIDS YIELDS ALL OF THE FOLLOWING, EXCEPT:
A. Phosphoric acid
B. Purines
C. 5-carbon sugar
D. Adenosine
E. Adenine
Ref. 1 - p. 181

928. RIBONUCLEIC ACID DOES NOT CONTAIN:
A. Adenine
B. Hydroxymethylcytosine
C. D-ribose
D. Uracil
E. Phosphate
Ref. 1 - p. 205

929. ALKALINE HYDROLYSIS (pH 9, 100° C) OF RNA YIELDS:
A. 2',3'-cyclic phosphates
B. Nucleoside-5'-phosphates
C. Nucleoside-3',5'-diphosphates
D. Nucleoside-2'-phosphates
E. Adenosine diphosphate Ref. 1 - p. 201

930. THE ALLOSTERIC SITE OF AN ENZYME:
A. Is simply another way of designating the active site
B. Binds molecules other than the substrate non-competitively, thus affecting enzyme activity
C. Is a misnomer and really implies three-point attachment of the substrate
D. Is a term derived from sterol metabolism and indicates a binding site for allo-compounds
E. Designates the site which catalyzes a reaction with the unnatural isomer of the substrate — Ref. 1 - pp. 242-246

931. A PHOSPHORIBOSYL TRANSFERASE CATALYZES THE FOLLOWING REACTION:
A. Purine + phosphoribosylpyrophosphate $\longrightarrow$ nucleotide
B. Purine + ATP $\longrightarrow$ nucleotide
C. Purine + ribose-1-phosphate $\longrightarrow$ nucleoside
D. All of these
E. None of these — Ref. 1 - p. 626

932. ADENYLIC ACID IS SYNTHESIZED BY REACTIONS INVOLVING:
A. Inosinic acid + NH_3
B. Inosinic acid + DPN + glutamine
C. Inosinic acid + guanosine triphosphate + aspartate
D. Hypoxanthine + ribose-1-phosphate
E. None of these — Ref. 1 - pp. 624-625

933. FOR NORMAL GROWTH YOUNG ANIMALS HAVE A REQUIREMENT FOR:
A. Purines
B. Pyrimidines
C. Both purines and pyrimidines
D. Nucleic acids
E. None of these — Ref. 1 - p. 619

934. THE FOUR NITROGEN ATOMS OF THE PURINE RING ARE DERIVED FROM:
A. Aspartate, glutamine and glycine
B. Glutamine, ammonia and aspartate
C. Glycine and aspartate
D. Ammonia, glycine and glutamate
E. Urea and ammonia — Ref. 1 - p. 620

935. THE FORMATION OF URIC ACID FROM PURINES IS CATALYZED BY:
A. Adenylic acid deaminase
B. Uricase
C. Allantoinase
D. Urease
E. Xanthine oxidase — Ref. 1 - p. 628

936. THE MAIN CATABOLIC PRODUCT OF PURINE METABOLISM IN MAN IS:
A. Allantoin
B. Urea
C. Ammonia
D. Uric acid
E. Hypoxanthine — Ref. 1 - p. 628

SECTION V - NUCLEOTIDES, PURINES, PYRIMIDINES

937. A SUGAR INHERENT IN THE STRUCTURE OF DNA IS:
A. L-2-Deoxyribofuranose
B. L-2-Deoxyribopyranose
C. D-2-Deoxyribopyranose
D. D-3-Deoxyribofuranose
E. D-2-Deoxyribofuranose
Ref. 1 - p. 184

938. THE FOLLOWING COMPOUND IS PRESENT IN RNA BUT ABSENT FROM DNA:
A. Thymine
B. Cytosine
C. Uracil
D. Guanine
E. Adenine
Ref. 1 - p. 202

939. NUCLEIC ACIDS CAN BE DETECTED BY MEANS OF THEIR ABSORPTION MAXIMA NEAR 260 nm. THEIR ABSORPTION IN THIS RANGE IS DUE TO:
A. Proteins
B. Purines and pyrimidines
C. Ribose
D. Deoxyribose
E. Phosphate
Ref. 1 - p. 188

940.

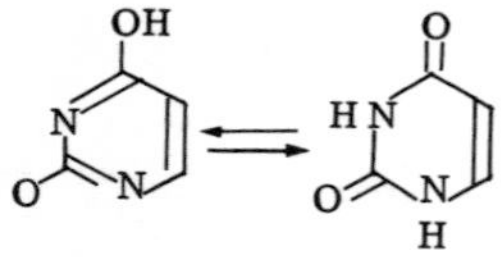

THESE STRUCTURES REPRESENT:
A. Optical isomerism
B. Geometric isomerism
C. Cis-trans isomerism
D. Lactam-lactim tautomerism
E. Purine-pyrimidine equilibrium
Ref. 1 - p. 183

941. HYDROLYTIC METHODS EXIST FOR THE STEPWISE DEGRADATION OF NUCLEOPROTEINS; LISTED IN DECREASING ORDER OF COMPLEXITY THE PRODUCTS OBTAINED ARE:
A. Nucleoprotein, nucleic acid, nucleotides, nucleosides, purines and pyrimidines
B. Nucleic acid, nucleoprotein, nucleotides, nucleosides, purines and pyrimidines
C. Nucleic acid, nucleoprotein, nucleosides, nucleotides, purines and pyrimidines
D. Purines and pyrimidines, nucleosides, nucleotides, nucleic acid, nucleoprotein
E. Nucleoprotein, nucleic acid, nucleosides, nucleotides, purines and pyrimidines
Ref. 1 - p. 181

942. THE RESISTANCE OF NUCLEOSIDES TO ALKALINE HYDROLYSIS IS DUE TO:
A. Their conversion to cyclic 2', 3'-monophosphates
B. The stability of ribose in alkali
C. The weakly acidic properties of purines
D. The stability of the beta-glycosidic linkage
E. The conversion to nucleotides Ref. 1 - p. 185

943. DIGESTION OF RIBONUCLEIC ACID BY PANCREATIC RIBONUCLEASE YIELDS:
A. A mixture of nucleoside 2'- and 3'-phosphates
B. Phosphodiesterases
C. Nucleoside-2'-phosphates
D. Nucleoside-5'-phosphates
E. Nucleoside-3'-phosphates Ref. 1 - p. 186

944. THE USUAL WAY OF DISTINGUISHING BETWEEN DNA AND RNA IS TO:
A. Determine their solubility in 1 M NaCl
B. Estimate the rate of trypsin digestion
C. Identify the sugar component colorimetrically
D. Extract the DNA with hot trichloroacetic acid
E. Measure their ultraviolet absorption spectra
Ref. 1 - p. 188

945. DNA IS:
A. Usually present in tissues as a nucleoprotein and cannot be separated from its protein component
B. A long chain polymer in which the internucleotide linkages are of the diester type between C-3' and C-5'
C. Different from RNA since in the latter the internucleotide linkages are between C-2' and C-5'
D. Hydrolyzed by weak alkali (pH 9 at 100° C)
E. Degraded by pancreatic deoxyribonuclease to yield mainly mono-nucleotide-5'-phosphates Ref. 1 - p. 189

946. THE PROBABLE METABOLIC DEFECT IN GOUT IS:
A. Hyperuricemia, a condition only found in patients with gout
B. An overproduction of uric acid due to lack of feedback control of phosphoribosylamine synthesis
C. An underproduction of purines leading to defective genetic material
D. Decreased rate of elimination of uric acid
E. Reduction of the miscible pool of uric acid
Ref. 1 - pp. 628-631

947. URIC ACID FORMATION IN MAMMALS TAKES PLACE IN:
A. Kidney
B. Liver
C. Intestine
D. Muscle
E. Liver and kidney Ref. 1 - p. 628

SECTION V - NUCLEOTIDES, PURINES, PYRIMIDINES

948. OROTIC ACID SYNTHESIS FROM ITS IMMEDIATE PRECURSOR INVOLVES:
A. Dihydro-orotase
B. Uridylic acid + FAD
C. An enzyme containing FAD, FMN and Fe
D. Glutamate + carbamyl phosphate
E. Uracil + DPN Ref. 1 - p. 634

949. AN ENZYME SYSTEM HAS BEEN ISOLATED FROM E. COLI WHICH CONVERTS DEOXYURIDINE-5'-PHOSPHATE TO THYMIDINE-5'-PHOSPHATE. WHICH OF THE FOLLOWING IS REQUIRED FOR THIS REACTION?:
A. Tetrahydrofolic acid
B. ATP
C. Flavin mononucleotide
D. 4'-phosphopantothenate
E. None of these Ref. 1 - p. 639

950. IN MOST MAMMALS EXCEPT MAN AND THE HIGHER APES, URIC ACID IS METABOLIZED BY:
A. Reduction to ammonia
B. Oxidation to allantoin
C. Degradation to allantoic acid
D. Hydrolysis to ammonia
E. Oxidation to ammonia and carbon dioxide
Ref. 1 - pp. 631-632

951. NUCLEOPROTEINS ARE CONJUGATED PROTEINS CONTAINING NUCLEIC ACID PLUS:
A. Albuminoid proteins
B. Globulins
C. Protamines or histones
D. Scleroproteins
E. Glycoproteins Ref. 1 - pp. 199-200

952. 5-PHOSPHORIBOSYL-1-PYROPHOSPHATE IS AN INTERMEDIATE IN THE SYNTHESIS OF:
A. Pyrimidines
B. Purine nucleotides
C. Pyrimidine nucleotides
D. DPN
E. All of the above
Ref. 1 - pp. 620-621

953. INOSINIC ACID IS THE BIOLOGICAL PRECURSOR OF:
A. Orotic acid and uridylic acid
B. Adenylic acid and guanylic acid
C. Purines and pyrimidines
D. Uracil and thymine
E. Uridylic acid and cytidylic acid Ref. 1 - pp. 624-625

954. IN THE BIOSYNTHESIS OF

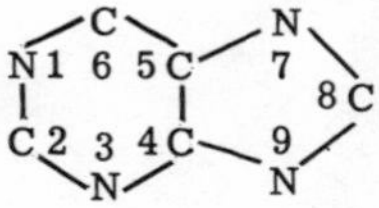

```
    C        N
N1 6 5C    7
|      |      8C
C2 3 4C    9
    N        N
```

, TRACER STUDIES HAVE SHOWN THAT:
A. The atoms in positions 7, 8, 9 are derived from urea
B. The atoms in positions 1, 6, 5 are derived from glycine
C. The atoms in positions 3 and 9 are derived from glutamine
D. The atom in position 9 is derived from aspartate
E. The atoms in position 3, 4, 5 are derived from glycine
Ref. 1 - p. 620

955. THE BASE COMPOSITION OF DNA FROM VARIOUS SOURCES HAS BEEN STUDIED IN SOME DETAIL. FIND THE STATEMENT BELOW WHICH IS INCORRECT:
A. In higher organisms, DNA from different tissues of the same species is similar in composition
B. The presence of methylcytosine is a characteristic feature of bacterial DNA
C. The base ratios A/T and G/C are usually close to unity
D. The ratio A + T/G + C ranges from 1.0–1.8 in mammalian DNA
E. A molecule of DNA may contain more than 20,000 nucleotides
Ref. 1 - pp. 191-192

956. THE STRUCTURAL STABILITY OF THE DOUBLE HELIX OF DNA IS ASCRIBED LARGELY TO:
A. Hydrogen bonding between adjacent purine bases
B. Hydrophobic bonding between stacked purine and pyrimidine nuclei
C. Hydrogen bonding between adjacent pyrimidine bases
D. The helical structure of the molecule
E. Hydrogen bonding between purine and pyrimidine bases
Ref. 1 - p. 193

ANSWER THE FOLLOWING GROUP OF QUESTIONS BY USING THE KEY OUTLINED BELOW:
A. If the item is associated with A only
B. If the item is associated with B only
C. If the item is associated with both A and B
D. If the item is associated with neither A nor B

A. DNA
B. RNA
C. Both
D. Neither

957. ___ Pentose attached to purine at position 9
958. ___ Pentose attached to pyrimidine at position 1
959. ___ Contains ribose in the furanose form
960. ___ Found in nearly all types of cells investigated
961. ___ Not affected by weak alkali Ref. 1 - pp. 184-185; 201

A. RNA
B. DNA
C. Both
D. Neither

962. ___ Contains D-2-deoxyribose
963. ___ Action of weak alkali yields as one of the products of hydrolysis cyclic nucleoside-2', 3'-phosphates
964. ___ Hydrolyzed by pancreatic ribonuclease
965. ___ Composed of 2 helical chains of polynucleotide held together by hydrogen bonding between pyrimidines and purines
966. ___ Contained almost exclusively in the cell nucleus

Ref. 1 - pp. 184; 193; 201

A. DNA
B. RNA
C. Both
D. Neither

967. ___ Contains the pyrimidine thymine
968. ___ Has strong ultraviolet absorption band at 260 nm
969. ___ Has strong ultraviolet absorption band at 280 nm
970. ___ Transforming factor
971. ___ Gives the Feulgen reaction for aldehydes

Ref. 1 - pp. 28; 186; 188-189; 192; 202

A. RNA
B. DNA
C. Both
D. Neither

972. ___ Internucleotide linkage of diester type between C-3' and C-5'
973. ___ Certain types contain pseudouridine
974. ___ Certain types contain ATP
975. ___ Denaturation produces a hyperchromic effect
976. ___ Sugar component determined colorimetrically

Ref. 1 - pp. 195-196; 200-202

THERE IS A WIDE SPECIES DIFFERENCE IN THE FINAL EXCRETORY PRODUCTS OF PURINE METABOLISM. MATCH THE DIFFERENT END PRODUCTS WITH THE GROUPS OF ANIMALS LISTED BELOW:

A. Uric acid
B. Allantoin
C. Allantoic acid
D. Urea
E. Ammonia

977. ___ Marine invertebrates
978. ___ Some teleost fishes
979. ___ Mammals (non-primate), turtles
980. ___ Man and other primates
981. ___ Most fishes, amphibia

Ref. 1 - p. 632

EACH GROUP OF QUESTIONS BELOW CONSISTS OF 5 LETTERED HEADINGS FOLLOWED BY A LIST OF NUMBERED WORDS OR PHRASES. FOR EACH NUMBERED WORD OR PHRASE SELECT THE ONE LETTERED HEADING THAT IS MOST CLOSELY RELATED TO IT:

A. Carbamyl phosphate
B. N^5, N^{10}-methylenetetrahydrofolate
C. Beta-alanine
D. Ammonia + ATP
E. ATP

982. ___ A substance required for synthesis of N-carbamyl aspartate, a precursor of pyrimidine nucleotides
983. ___ Required for conversion of uridine triphosphate to cytidine triphosphate
984. ___ The methyl group of thymine is derived from formate or from the 1-carbon pool. _________ is required for its addition to deoxyuridine -5'-phosphate to form thymidine-5'-phosphate
985. ___ Required for synthesis of deoxyribonucleotide triphosphates
986. ___ A metabolite of cytosine

Ref. 1 - pp. 633-640

A. 6-aminopurine
B. 6-oxypurine
C. 2, 6, 8-trioxypurine
D. 2, 4-dioxypyrimidine
E. 5-methyl-2, 4-dioxypyrimidine

987. ___ Thymine
988. ___ Uracil
989. ___ Hypoxanthine
990. ___ Adenine
991. ___ Uric acid

Ref. 1 - pp. 182-183

A. Azaserine
B. Aspartate
C. Alpha-N-formylglycinamide ribonucleotide
D. Glutamine + ATP + Mg^{++}
E. Inosinic acid

992. ___ A competitive inhibitor of reaction between 5-phosphoribosyl-1-pyrophosphate and glutamine to form 5-phosphoribosyl-1-amine, which thereby inhibits growth of certain neoplasms
993. ___ An intermediate in purine synthesis, which requires a folic acid derivative for its formation
994. ___ The first purine derivative synthesized de novo, which contains a completed ring structure
995. ___ Source of the amino group of adenylic acid
996. ___ Required for synthesis of guanylic acid from xanthylic acid

Ref. 1 - pp. 621-622, 624-625

A. Purine
B. Pyrimidine
C. Nucleoside
D. Nucleotide
E. Nucleic acid

997. ___ Adenylic acid
998. ___ Adenosine
999. ___ Adenine
1000. ___ Uracil
1001. ___ DNA
1002. ___ RNA

Ref. 1 - pp. 181-187

A.
H
C
N CH
HC CH
N

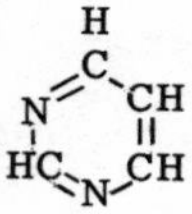

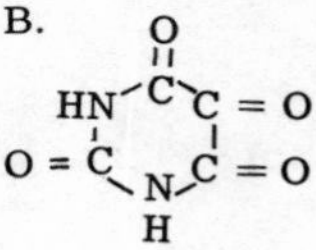

B.
O
C
HN C = O
O = C C = O
N
H

C.
H
C C N
N CH
HC N C N
H

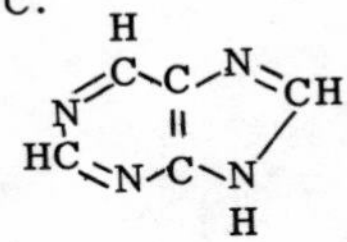

D.
OH
C
N C $-CH_3$
C CH
HO N

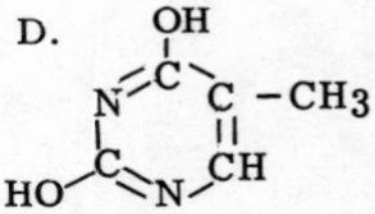

E.
OH
C C N
N COH
C N C N
HO H

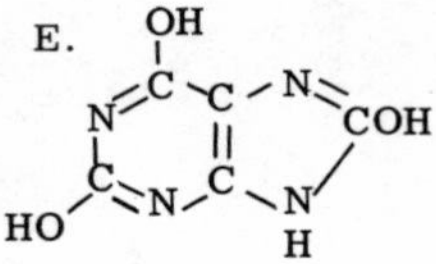

1003. ___ Alloxan
1004. ___ Purine
1005. ___ Uric acid
1006. ___ Pyrimidine
1007. ___ Thymine

Ref. 1 - pp. 182-184

A.

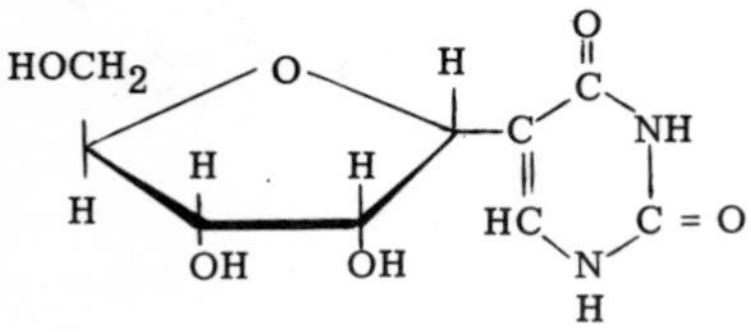

B.

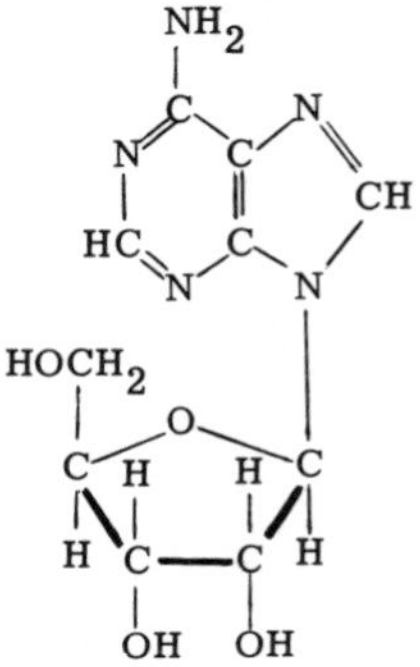

C.

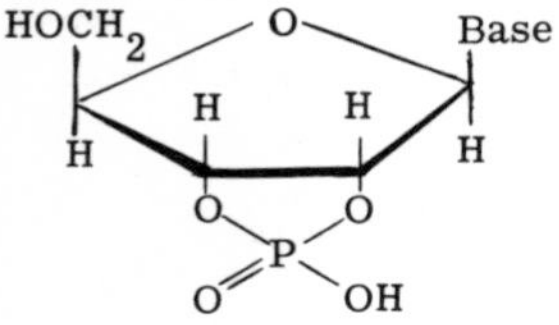

D.

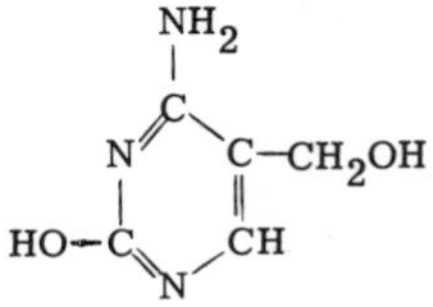

E.

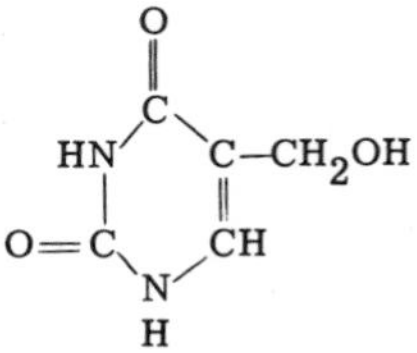

1008. ___ 2', 3' cyclic monophosphate nucleotide
1009. ___ Adenosine
1010. ___ 5-Hydroxymethyluracil
1011. ___ Pseudouridine
1012. ___ 5-Hydroxymethylcytosine

Ref. 1 - pp. 185, 201-205

MATCH THE FORMULAS WITH THE CORRECT NAME:

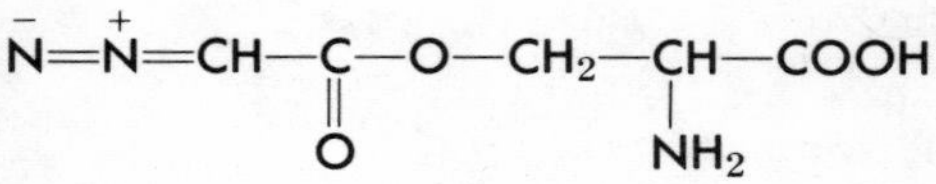

A.

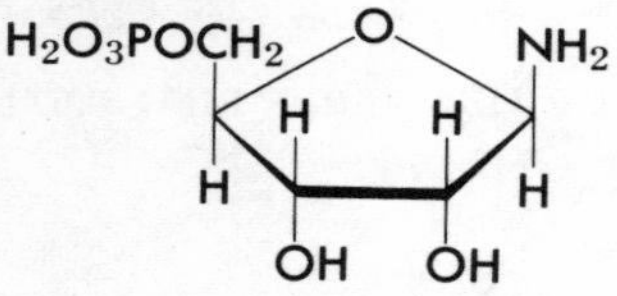

B.

+

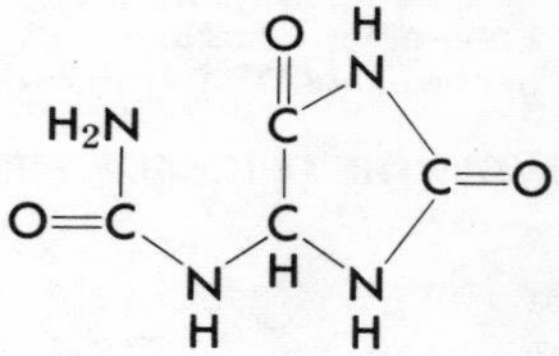

C.

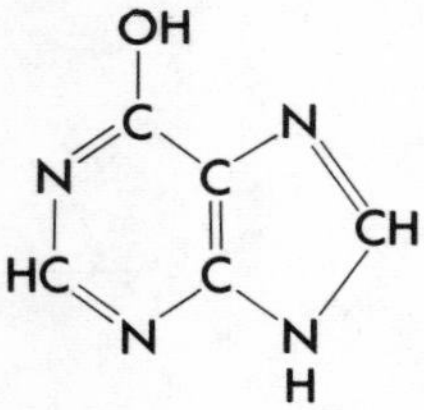

D.

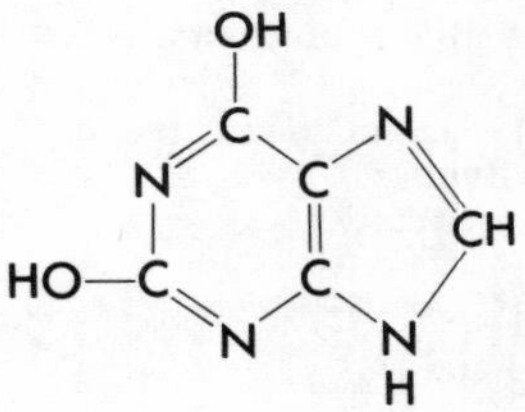

E.

1013. ___ 5-Phosphoribosyl-1-amine

1014. ___ Xanthine

1015. ___ Allantoin

1016. ___ Azaserine

1017. ___ Hypoxanthine

Ref. 1 - pp. 621; 629

RELATE EACH OF THE FOLLOWING METABOLIC INHIBITORS TO THE REACTION OR REACTION SEQUENCE IN WHICH IT IS KNOWN TO FUNCTION:

A. Mitomycin C
B. 6-mercaptopurine
C. 6-azauracil
D. 5-fluorodeoxyuridine
E. Methotrexate

1018. ___ Inosinic acid ⟶ adenylic acid
1019. ___ Decarboxylation of orotidylic acid
1020. ___ Synthesis of thymidylate
1021. ___ Folic acid reduction
1022. ___ Inhibition of DNA synthesis

Ref. 3 - pp. 232, 840-842

WHAT IS THE ORIGIN OF THE ATOMS OF THE PURINE RING SYSTEM?:

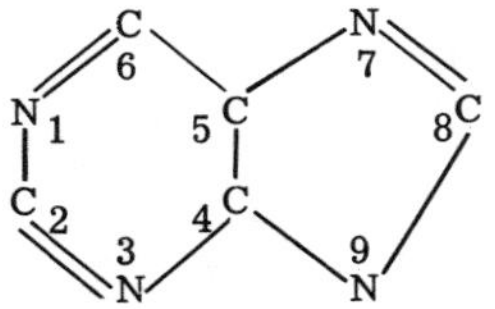

A. Amide N of glutamine
B. CO_2
C. 1-carbon donors (serine, glycine)
D. Glycine
E. Aspartate

1023. ___ N_1
1024. ___ N_3 and N_9
1025. ___ C_6
1026. ___ C_2 and C_8
1027. ___ N_7

Ref. 1 - p. 620

EACH OF THE QUESTIONS OR INCOMPLETE STATEMENTS BELOW IS FOLLOWED BY 5 SUGGESTED ANSWERS OR COMPLETIONS. SELECT THE ONE WHICH IS BEST IN EACH CASE:

1028. THE "COMMITTED" METABOLIC STEP IN PURINE BIOSYNTHESIS IS PROBABLY:
A. Ring closure of allantoin to reform uric acid
B. The reaction between 5-phosphoribosyl-1-pyrophosphate + glutamine $\longrightarrow$ 5-phosphoribosyl-1-amine + glutamic acid
C. The transformation of ribose-5-phosphate to 5-phosphoribosyl-1-pyrophosphate
D. The formation of an azaserine antagonist which prevents glutamine metabolism
E. Ring closure of formamidoimidazole-carboxamide ribonucleotide yielding inosinic acid Ref. 1 - p. 621

1029. EARLY STUDIES BY BUCHANAN SHOWED THAT ONE OF THE CARBONS OF THE PURINE RING SYSTEM WAS DERIVED FROM CO_2. THE PRECURSOR PARTICIPATING IN CO_2 FIXATION IS:
A. Glycinamide ribonucleotide
B. Inosinic acid
C. Aspartate
D. Alpha-N-formylglucinamidine ribonucleotide
E. 5-aminoimidazole ribonucleotide Ref. 1 - p. 623

1030. IN PURINE BIOSYNTHESIS 2 STEPS REQUIRE THE PRESENCE OF TETRAHYDROFOLATE. THESE ARE:
A. Formation of phosphoribosylamine and glycinamide ribonucleotide
B. Formation of alpha-N-formylglycinamide ribonucleotide and 5-amino-imidazole-4-carboxamide ribonucleotide
C. CO_2-fixation yielding 5-aminoimidazole-4-carboxylic acid ribonucleotide and formation of inosinic acid
D. Formation of alpha-N-formylglycinamide ribonucleotide and 5-formamidoimidazole-4-carboxamide ribonucleotide
E. Formation of inosinic acid and adenylosuccinic acid
Ref. 1 - pp. 622-624

1031. A KEY SUBSTANCE IN PYRIMIDINE BIOSYNTHESIS IS:
A. Carbamyl phosphate
B. Thiouracil
C. ATP
D. TPN^+
E. Ribose-5-phosphate Ref. 1 - p. 633

1032. THE "COMMITTED STEP" IN PYRIMIDINE BIOSYNTHESIS:
A. Provides an excellent example of positive feedback
B. Results in the formation of dihydro-orotic acid
C. Is the formation of N-carbamylaspartic acid
D. Is catalyzed by orotic decarboxylase
E. Requires ATP Ref. 1 - pp. 244; 633

1033. WHICH OF THE FOLLOWING STATEMENTS CONCERNING NAD AND NADP IS MOST NEARLY CORRECT ?:
A. NAD and NADP are similar chemically except that NAD has an additional phosphoryl group at position 2 on the pentose adjacent to adenine
B. NADP has an addition phosphoryl group at position 2 on the pentose adjacent to nicotinamide
C. NADP has an additional phosphoryl group at position 2 on the pentose adjacent to adenine
D. NAD and NADP differ only with respect to their redox potential
E. NAD is a hydride ion acceptor while NADP is a proton acceptor
Ref. 3 - p. 410

ANSWER THE FOLLOWING GROUP OF QUESTIONS BY USING THE KEY OUTLINED BELOW:
A. If the item is associated with A only
B. If the item is associated with B only
C. If the item is associated with both A and B
D. If the item is associated with neither A nor B

A. ATP (adenosine triphosphate)
B. UTP (uridine triphosphate)
C. Both
D. Neither

1034. ___ Used in "activation" of amino acids for protein synthesis
1035. ___ Used in sulfate "activation"
1036. ___ Used in synthesis of malonyl-CoA
1037. ___ Used in synthesis of FAD
1038. ___ Used in synthesis of a constituent of hyaluronic acid : N-acetyl-glucosamine
1039. ___ Used in the formation of CTP (cytidine triphosphate)
Ref. 1 - pp. 434; 487; 563-564; 635; 641; 662

A. ATP
B. UTP
C. Both
D. Neither

1040. ___ Involved in formation of TPN from DPN
1041. ___ Involved in glycogen synthesis by formation of a nucleotide-glucose complex
1042. ___ Involved in formation of a nucleotide-choline complex required in phospholipid synthesis
1043. ___ Involved in formation of urinary glucuronides
1044. ___ Involved in thiophorase reaction on succinyl-CoA
Ref. 1 - pp. 433; 443; 480; 509; 642

ANSWER THE FOLLOWING QUESTIONS BY USING THE KEY OUTLINED BELOW:
A. If A is greater than B
B. If B is greater than A
C. If A and B are equal or nearly equal

1045. A. Purine content of milk, cheese and eggs
B. Purine content of liver
Ref. 1 - p. 630

1046. A. Entry of protein of bacteriophage into host cell
B. Entry of DNA of bacteriophage into host cell
Ref. 1 - p. 205

1047. A. Nicotinamide content of DPN
B. Nicotinamide content of TPN
Ref. 1 - pp. 641-642

1048. A. Orotic acid incorporation into tissue purines
B. Orotic acid incorporation into tissue pyrimidines
Ref. 1 - p. 634

1049. A. Occurrence of guanine gout in pigs
B. Occurrence of uric acid gout in man
Ref. 1 - pp. 628-630

1050. A. Allantoin excretion in mammals (except primates)
B. Allantoin excretion in birds
Ref. 3 - p. 837

1051. A. Uricase activity in liver of birds
B. Uricase activity in liver of mammals (except primates)
Ref. 3 - p. 837

1052. A. Occurrence of gout in males
B. Occurrence of gout in females
Ref. 1 - p. 630

1053. A. Inhibition of xanthine oxidase by allopurinol
B. Inhibition of xanthine oxidase by alloxanthine
Ref. 1 - p. 631

1054. A. Concentration of uricase in man
B. Concentration of uricase in other primates
Ref. 1 - p. 628

1055. A. Occurrence of guanine gout in man
B. Occurrence of guanine gout in swine
Ref. 1 - p. 628

1056. A. Overproduction of uric acid in gout
B. Decrease in elimination of uric acid in gout
Ref. 1 - p. 631

1057. A. Participation of carbamyl phosphate in purine synthesis
B. Participation of orotic acid in pyrimidine synthesis
Ref. 1 - pp. 633-634

1058. A. Excretion of nitrogen as urea by uricotelic animals
B. Excretion of nitrogen as uric acid by ureotelic animals
Ref. 1 - p. 632

1059. A. Miscible body pool of uric acid in gouty subjects
B. Miscible body pool of uric acid in normal subjects
Ref. 1 - p. 630

ANSWER THE FOLLOWING QUESTIONS BY USING THE KEY OUTLINED BELOW:
A. If 1, 2 and 3 are correct
B. If 1 and 3 are correct
C. If 2 and 4 are correct
D. If only 4 is correct
E. If all four are correct

1060.
1. The enzyme DPNase is involved in the conversion of DPN to TPN
2. The enzyme DPNase catalyzes the hydrolysis of DPN at the N-glycosidic linkage
3. The enzyme DPNase hydrolyzes adenosine diphosphate ribose to adenylic acid and ribose-5-phosphate
4. The enzyme nucleotide pyrophosphatase catalyzes the conversion of DPN to nicotinamide mononucleotide + AMP

Ref. 1 - p. 642

1061. THYMINE:
1. Is a constituent of DNA
2. Has a methyl group which is derived from creatine
3. Has a methyl group which can be derived from methylenetetrahydrofolate
4. Introduction of the methyl group of thymine involves the transfer of a C_1 unit from a thiamine compound

Ref. 1 - pp. 182, 639

1062. OROTIC ACID:
1. Reacts with 5-phosphoribosyl-1-pyrophosphate to form orotidine-5'-phosphate
2. Is utilized for pyrimidine synthesis
3. Can be converted enzymatically to uridine and cytidine triphosphates in a series of reactions
4. Is a precursor in the formation of adenine nucleotides

Ref. 1 - pp. 634-735

1063. OROTIC ACID ACIDURIA:
1. Is enhanced by the administration of uracil or cytosine
2. Is a hereditary disorder of purine metabolism in man
3. Is associated with a block in the reduction of L-dihydroorotic acid
4. Is a hereditary disorder of pyrimidine metabolism in man

Ref. 1 - p. 635

1064. IN THE FORMATION OF PURINE AND PYRIMIDINE NUCLEOTIDES:
1. The pathways involved are not identical
2. All intermediates of the pyrimidine nucleotides are derivatives of ribose-5-phosphate
3. All intermediates of purine nucleotides are derivatives of ribose-5-phosphate
4. 5-phosphoribosyl-1-pyrophosphate is not involved in the synthesis of pyrimidine nucleotides Ref. 1 - pp. 620-621

1065. 1. In the transformation of deoxyuridine 5-phosphate to thymidine 5'-phosphate tetrahydrofolate serves both as carbon carrier and as hydrogen donor
2. In the conversion of cytidine diphosphate to deoxycytidine diphosphate a direct dehydrogenation is involved
3. In the conversion of cytidine diphosphate to deoxycytidine diphosphate an unsaturated derivative of ribose is an intermediate
4. Ribonucleotides cannot be converted directly to deoxyribonucleotides; i.e. deoxygenation of ribose cannot occur at the nucleotide level
Ref. 1 - p. 639

EACH OF THE QUESTIONS OR INCOMPLETE STATEMENTS BELOW IS FOLLOWED BY 5 SUGGESTED ANSWERS OR COMPLETIONS. SELECT THE ONE WHICH IS BEST IN EACH CASE:

1066. THE DNA MOLECULE CONSISTS OF TWO POLYNUCLEOTIDE CHAINS. THESE CHAINS:
A. Are held together by phosphate groups
B. Are connected by disulfide bonds
C. Are identical in chemical composition
D. Differ in chemical composition
E. Cannot be separated

Ref. 1 - pp. 193, 649
Ref. 4 - pp. 124-126

1067. THE TRANSFORMING PRINCIPLE OF BACTERIA HAS BEEN IDENTIFIED AS:
A. RNA
B. DNA
C. sRNA
D. tRNA
E. mRNA

Ref. 2 - p. 50
Ref. 4 - p. 126

1068. mRNA:
A. Is identical to sRNA
B. Contains the information for protein synthesis
C. Must exist in at least 20 different forms, one for each amino acid
D. Is sometimes called a cistron
E. Imparts no information regarding protein synthesis

Ref. 1 - p. 661

1069. tRNA:
A. Is identical to mRNA
B. Must exist in at least 20 different forms, one for each amino acid
C. Contains the information for protein synthesis
D. Is sometimes called a muton
E. Contains thymidine

Ref. 1 - pp. 661-664

1070. THE GENETIC INFORMATION OF NUCLEAR DNA MUST BE TRANSMITTED TO THE SITE OF PROTEIN SYNTHESIS. THIS IS DONE BY:
A. rRNA
B. sRNA
C. DNA, directly
D. Polysomes
E. mRNA

Ref. 1 - p. 661

1071. mRNA IS A COMPLEMENTARY COPY OF:
A. A single strand of DNA
B. Both strands of DNA
C. Ribosomal DNA
D. rRNA
E. sRNA

Ref. 1 - p. 661

1072. THE SPECIFIC AMINO ACID SEQUENCE IN A POLYPEPTIDE IS FUNDAMENTALLY DETERMINED BY:
A. Conformation of ribosomal protein
B. The base sequence of mRNA
C. The base sequence of DNA
D. Double stranded RNA
E. The base sequence of rRNA

Ref. 1 - p. 661
Ref. 4 - pp. 130-132

1073. IN MAMMALIAN CELLS rRNA IS PRODUCED MAINLY IN THE:
A. Endoplasmic reticulum
B. Nucleolus
C. Lysosome
D. Ribosome
E. Polysome
Ref. 2 - p. 54

1074. IN MAMMALIAN CELLS PROTEIN SYNTHESIS TAKES PLACE PREDOMINANTLY ON THE:
A. Endoplasmic reticulum
B. Nucleolus
C. Lysosome
D. Ribosome
E. Polysome
Ref. 2 - p. 96

1075. IN PROTEIN BIOSYNTHESIS:
A. Each amino acid "recognizes" its codon on the mRNA template because of structural specificity
B. Exactness of read-out is assured by the presence of traces of DNA on the ribosome
C. Each amino acid is first attached to an anticodon specific for the amino acid
D. A given codon - anticodon pair must have identical base sequences to avoid the formation of "degenerate" proteins
E. Each amino acid "recognizes" its codon through recognition nucleotides in its specific transfer RNA molecule
Ref. 2 - p. 94

ANSWER THE FOLLOWING QUESTION BY USING THE KEY OUTLINED BELOW:
A. If 1, 2 and 3 are correct
B. If 1 and 3 are correct
C. If 2 and 4 are correct
D. If only 4 is correct
E. If all four are correct

1076. THE MECHANISM OF INDUCTION AND REPRESSION POSTULATED BY JACOB AND MONOD IMPLIES THAT:
1. Both induction and repression operate by affecting protein synthesis
2. There are structural genes and regulatory genes
3. Several structural genes are active or inactive simultaneously
4. These phenomena do not function at the genetic level
Ref. 1 - p. 685

FOR EACH OF THE FOLLOWING MULTIPLE CHOICE QUESTIONS SELECT THE ONE MOST APPROPRIATE ANSWER:

1077. THE MOST ACTIVE SITE OF PROTEIN SYNTHESIS IS THE:
A. Nucleus
B. Ribosome
C. Mitochondrion
D. Cell sap
E. Lysosome
Ref. 2 - p. 96

1078. IN THE WATSON-CRICK MODEL OF DNA:
A. The 2 chains are identical
B. If one chain forms a right-handed helix the other must be left-handed
C. The structure was deduced by X-ray diffraction analysis
D. Hydrogen bonding is too weak to account for the stability of the molecule
E. The ratio of $\frac{A + T}{G + C}$ must equal 1
Ref. 1 - pp. 193-195

1079. THE INFORMATION CONTAINED IN NUCLEAR DNA IS TRANSMITTED TO THE SITES OF PROTEIN SYNTHESIS BY:
A. DNA fragments which leave the nucleus
B. Ribosomes entering the nucleus
C. Transfer RNA
D. Messenger DNA
E. Messenger RNA
Ref. 2 - p. 130

1080. WHICH OF THE FOLLOWING STATEMENTS CONCERNING THE BASE COMPOSITION OF DNA IS INCORRECT?:
A. The base composition of DNA is characteristic of a given species
B. Closely related organisms exhibit similar base composition
C. If $A + G + C + T \neq 1$, the DNA involved may be of the non-Watson-Crick type
D. $A + T/G + C$ has been defined as the dissymmetry ratio
E. Different tissues of the same organism have different base composition
Ref. 3 - pp. 202-203

1081. IN THE BIOSYNTHESIS OF ø X174 DNA THE JOINING ENZYME (LIGASE) REQUIRES:
A. UTP
B. FMN
C. DPN
D. TPN
E. ATP
Ref. 1 - p. 654

1082. PROTEIN SYNTHESIS IN THE CYTOPLASM OF INTACT RAT LIVER CELLS IS NOT AFFECTED BY ONE OF THE FOLLOWING:
A. Actinomycin
B. Puromycin
C. Cycloheximide
D. Chloramphenicol
E. Ethionine
Ref. 1 - p. 670

1083. WHICH TERM IS USED TO DESCRIBE A MUTATION OF ONE GENE OF AN OPERON CAUSING DECREASED OPERATION OF THE GENES DISTAL TO THE OPERATOR:
A. Negative feedback
B. Feedback repression
C. Dilution
D. Deletion
E. Modulation
Ref. 1 - pp. 682-683

1084. WHICH OF THE FOLLOWING STATEMENTS CONCERNING RIBOSOMES IS INCORRECT?:
A. The order of nucleotides in rRNA carries no coding for amino acids
B. Ribosomes consist of approximately equal weights of protein and RNA
C. The composition of rRNA is not directly specified by DNA
D. rRNA is made as a single chain in the nucleolus and then hydrolyzed into at least 3 pieces
E. One of the protein components of the ribosomes is necessary to attach incoming tRNA to the aminoacyl tRNA site
Ref. 5 - pp. 33-34

1085. THE FUNDAMENTAL GENETIC IMPORTANCE OF DNA WAS FIRST INDICATED BY THE FOLLOWING OBSERVATION:
A. All mammalian cells contain DNA
B. DNA forms a double helix
C. Transforming factor is DNA
D. All viruses contain DNA or circular RNA
E. Some DNA is extranuclear
Ref. 2 - p. 50
Ref. 4 - p. 126

1086. HISTONES ARE DEFINED AS:
A. Acidic proteins associated with DNA
B. Acidic proteins associated with RNA
C. Neutral proteins associated with DNA
D. Basic proteins associated with DNA
E. Basic proteins associated with RNA
Ref. 2 - p. 30
Ref. 3 - p. 256

1087. A CERTAIN TYPE OF RNA CHARACTERISTICALLY CONTAINS METHYLATED PURINES AND PYRIMIDINES. THE RNA IN QUESTION IS:
A. tRNA
B. mRNA
C. rRNA
D. Viral RNA
E. Phage RNA
Ref. 2 - pp. 43-44

1088. CODON CONSISTS OF:
A. One molecule of aminoacyl-tRNA
B. 2 complementary base pairs
C. 3 consecutive nucleotide units
D. An individual ribosome
E. 4 individual nucleotides
Ref. 2 - pp. 92-93

1089. tRNA REACTS SPECIFICALLY WITH:
A. Specific aminoacyl adenylates
B. ATP
C. Free amino acids
D. The Golgi apparatus
E. Nuclear RNA
Ref. 2 - p. 94

1090. RIBOSOMAL RNA:
A. Is complementary to nuclear DNA
B. Accounts for the greater part of all the RNA present in the cell
C. Has the same molecular weight as mRNA
D. Is identical with tRNA
E. Contains the amino acid activating enzyme
Ref. 2 - p. 43

1091. THE PHENOMENA OF INDUCTION AND REPRESSION INDICATE THAT:
A. Constitutive enzymes are under the influence of inducers
B. There probably exist structural genes and regulatory genes
C. The genes for enzyme synthesis can be produced as needed
D. There exists one inducer for a given enzyme
E. Derepression involves the conversion of a zymogen to an enzyme
Ref. 1 - pp. 684-685

1092. THE ISOZYMES OF LDH:
A. Demonstrate the evolutionary pathway of this enzyme
B. Range from monomers to tetramers
C. Differ only in a single amino acid
D. Exist in 5 forms depending upon the ratio of A monomer to B monomer
E. Are forms of the enzyme which differ in activity but not in electrophoretic mobility
Ref. 2 - pp. 130-131
Ref. 3 - p. 174

1093. TOTAL RIBOSOMAL RNA OF RAT LIVER IS CHARACTERIZED BY HIGH PROPORTIONS OF:

A. Adenine + Cytosine
B. Guanine + Cytosine
C. Adenine + Guanine
D. Guanine + Uracil
E. Uracil + Adenine

Ref. 3 - p. 247

1094. HISTONES:

A. Are basic proteins rich in lysine and/or arginine
B. Are bound covalently to DNA
C. Have relatively high molecular weights (200,000 or higher)
D. Are identical with protamines
E. Cannot be dissociated intact from DNA

Ref. 3 - pp. 256-258

1095. WHICH OF THE FOLLOWING STATEMENTS CONCERNING MESSENGER RNA IS INCORRECT ?:

A. The sugar moiety of mRNA is D-ribose
B. The polynucleotide chain of mRNA is longer than that of DNA
C. mRNA exists as single-stranded molecules
D. The codons in mRNA are the complements of the codons in the active strand of DNA
E. The complement of a mRNA codon is referred to as an anticodon

Ref. 5 - pp. 31-32

ANSWER THE FOLLOWING QUESTIONS BY USING THE KEY OUTLINED BELOW:

A. If the item is associated with A only
B. If the item is associated with B only
C. If the item is associated with both A and B
D. If the item is associated with neither A nor B

A. mRNA
B. tRNA
C. Both
D. Neither

1096. ___ Composition strongly dependent upon Mg^{++} concentration
1097. ___ Transmits genetic information from nuclear DNA to sites of protein synthesis
1098. ___ React(s) specifically with a single amino acid
1099. ___ Molecular weight about 30,000
1100. ___ Contain(s) thymidine
1101. ___ Contain(s) uracil

Ref. 1 - pp. 201, 661-664
Ref. 4 - pp. 128-133

A. DNA
B. mRNA
C. Both
D. Neither

1102. ___ Carrier of amino acids to site of protein synthesis
1103. ___ Made in nucleus of cell
1104. ___ Possess(es) property of self-replication
1105. ___ Synthesis catalyzed by DNA-dependent RNA-polymerase
1106. ___ Helical structure
1107. ___ Produced by structural gene

Ref. 1 - pp. 661; 666-667;685
Ref. 4 - pp. 128-133

EACH OF THE QUESTIONS BELOW CONSISTS OF 5 LETTERED HEADINGS FOLLOWED BY A LIST OF NUMBERED WORDS OR PHRASES. FOR EACH NUMBERED WORD OR PHRASE SELECT THE ONE LETTERED HEADING THAT IS MOST CLOSELY RELATED TO IT:

RELATE THE FOLLOWING INBORN ERRORS OF METABOLISM TO THE LACK OR MODIFICATION OF AN ENZYME OR PROTEIN:

A. Glucose-6-phosphatase
B. Iodotyrosine dehalogenase
C. Ceruloplasmin
D. Tryptophan pyrrolase
E. Phenylalanine hydroxylase

1108. ___ Goiter
1109. ___ Hartnup's disease
1110. ___ Von Gierke's disease
1111. ___ Phenylketonuria
1112. ___ Wilson's disease

Ref. 1 - p. 700

MATCH THE INHIBITORS OF PROTEIN SYNTHESIS WITH THEIR POSTULATED MODES OF ACTION:

A. Puromycin
B. Tetracycline
C. Aurintricarboxylic acid
D. Cycloheximide
E. Sparsomycin

1113. ___ Blocks aminoacyl tRNA site
1114. ___ Binds to 60S ribosome
1115. ___ Interferes with peptide bond formation
1116. ___ Interferes with attachment of mRNA
1117. ___ Induces premature chain termination

Ref. 3 - pp. 954-955

A. Degeneracy of code
B. Comma-less code
C. Overlapping code
D. Coding ratio
E. Universality of code

1118. ___ Number of nucleotides per codon
1119. ___ Amino acids are coded for by more than one codon
1120. ___ Elements of certain code groups form parts of other code groups
1121. ___ Code is in use throughout living things
1122. ___ No spacings between adjacent code groups

Ref. 3 - pp. 869-874

ARRANGE THE FOLLOWING STEPS OF POLYPEPTIDE SYNTHESIS IN THEIR CORRECT SEQUENCE:

A. Formation of complex between ribosomes and mRNA
B. Formation of aminoacyl-tRNAs
C. Release of polypeptide chain
D. Activation of amino acid
E. Formation of peptide bonds

1123. ___ First step
1124. ___ Second step
1125. ___ Third step
1126. ___ Fourth step
1127. ___ Fifth step

Ref. 3 - p. 918

MATCH THE HEREDITARY DISORDER WITH THE AFFECTED ENZYME OR PROTEIN:

A. Histidine decarboxylase
B. β-Lipoprotein
C. p-Hydroxyphenylpyruvic
D. Accelerator globulin
E. Iodotyrosine dehalogenase

1128. ___ Acanthocytosis
1129. ___ Familial goiter
1130. ___ Mastocytosis
1131. ___ Parahemophilia
1132. ___ Tyrosinosis

Ref. 3 - p. 857

SIX OR SEVEN TYPES OF GLYCOGEN STORAGE DISEASE ARE CURRENTLY RECOGNIZED. RELATE THE ENZYMIC DEFECT WITH THE TYPE (AND NAME) OF THE DISEASE:

A. Type I (von Gierke's disease)
B. Type II (Pompe's disease)
C. Type III (Forbes' disease)
D. Type IV (Andersen's disease)
E. Type V (McArdle's disease)

1133. ___ Amylo-α-1, 4-glucosidase
1134. ___ Amylo-(1, 4 → 1, 6) transglucosylase
1135. ___ Glucose-6-phosphatase
1136. ___ Muscle glycogen phosphorylase
1137. ___ α-1, 4 → 1, 4-glucan transferase Ref. 1 - pp. 446, 884
[see also Ref. 5 - pp. 305-308]

MATCH THE FOLLOWING INHIBITORS OF PROTEIN SYNTHESIS WITH THEIR POSTULATED MODE OF ACTION:

A. 5-Fluorouracil
B. 5-Methyltryptophan
C. Chloramphenicol
D. Actidione
E. Puromycin

1138. ___ Makes functional ribosomes unavailable in bacteria
1139. ___ Functions as analog of amino acylated tRNA
1140. ___ Is incorporated into RNA
1141. ___ Inhibits amino acyl synthetase of parent compound
1142. ___ Makes functional ribosomes unavailable in higher organisms
Ref. 3 - pp. 188, 953-957

RELATE THE FOLLOWING HEREDITARY DISORDERS TO THE DEFICIENCY OR ABSENCE OF A SPECIFIC PROTEIN:

A. Phenylketonuria
B. Wilson's disease
C. Maple syrup urine disease
D. Crigler-Najjar syndrome
E. Hemolytic anemia

1143. ___ UDP-glucuronate transferase
1144. ___ Ceruloplasmin
1145. ___ Amino acid decarboxylase
1146. ___ Pyruvate kinase
1147. ___ Phenylalanine hydroxylase Ref. 2 - p. 102

RELATE THE FOLLOWING HEREDITARY DISORDERS TO THE LACK OF A SPECIFIC PROTEIN:

A. Hartnup's disease
B. Agammaglobulinemia
C. Parahemophilia
D. Analbuminemia
E. Acanthocytosis

1148. ___ Accelerator globulin
1149. ___ Serum albumin
1150. ___ Low density lipoprotein
1151. ___ Gamma-globulin
1152. ___ Tryptophanpyrrolase Ref. 1 - p. 700

RELATE THE FOLLOWING HEREDITARY DISORDERS TO THEIR BIOCHEMICAL MANIFESTATION:

A. Cystinosis
B. Cystinuria
C. Congenital steatorrhea
D. Hurler's syndrome
E. Porphyria

1153. ___ Aberration of amino acid transport into cells
1154. ___ Excretion of chondroitin sulfate B
1155. ___ Excretion of uroporphyrins
1156. ___ Excretion of cystine, lysine, etc.
1157. ___ Failure to absorb lipid Ref. 1 - p. 701

MATCH THE FOLLOWING:

A. Transfer enzyme
B. DNA polymerase
C. Amino acyl synthetase
D. Polynucleotide phosphorylase
E. DNA-dependent RNA polymerase

1158. ___ Catalyzes first step of protein synthesis
1159. ___ Catalyzes formation of mRNA
1160. ___ Catalyzes attachment of amino acids to ribosome
1161. ___ Useful in synthesis of synthetic mRNA. Does not require DNA primer
1162. ___ Requires single-stranded DNA primer

Ref. 1 - pp. 650; 662; 666-669; 672-673

ANSWER THE FOLLOWING QUESTIONS BY USING THE KEY BELOW:
A. If 1, 2 and 3 are correct
B. If 1 and 3 are correct
C. If 2 and 4 are correct
D. If only 4 is correct
E. If all four are correct

1163. mRNA:
1. Has a helical, double stranded structure
2. Must be capable of diffusing from the nucleus into the cytoplasm
3. Its biosynthesis requires sRNA as a primer
4. In certain viruses mRNA must be the genetic material since DNA is absent
Ref. 1 - p. 671
Ref. 4 - pp. 130-133

1164. PROTEIN SYNTHESIS:
1. Requires at least one specific activating enzyme for each different amino acid incorporated
2. The specific amino acid sequence is controlled by rRNA
3. Takes place on the ribosome
4. Can be induced only by other proteins Ref. 4 - pp. 128-133, 139

1165. INBORN ERRORS OF METABOLISM:
1. May be due to the absence of a final product of a metabolic pathway
2. May be due to an accumulation of a non-metabolizable intermediate
3. May be due to the accumulation of products of an ordinarily minor metabolic pathway
4. May be due to a mutation which produces an entirely new metabolic pathway Ref. 1 - pp. 699-701

1166. IN THE BIOSYNTHESIS OF HUMAN HEMOGLOBIN:
1. The biosynthesis of the alpha and beta chains is controlled by a single gene
2. The amino acid sequence is determined by the sequence of nucleotides in DNA
3. The specific folding is due to the specific location of disulfide bonds
4. An association of four beta-chains can occur in rare cases
Ref. 1 - pp. 657-658

1167. rRNA:
1. Is another name for polysome
2. Is a lipoprotein since it is deactivated by deoxycholate
3. Is destroyed by treatment with ribonuclease
4. Is located on ribosomes and polysomes
Ref. 1 - pp. 664-665

ANSWER THE FOLLOWING QUESTIONS BY USING THE KEY BELOW:
A. If 1 is correct
B. If 2 is correct
C. If both 1 and 2 are correct
D. If both 1 and 2 are incorrect

1168. HYDROLYSIS STUDIES ON ADENYLIC ACID HAVE SHOWN THAT:
1. Mild acid hydrolysis separates the phosphate from the nucleotide
2. Mild alkaline hydrolysis splits off the pentose phosphate
Ref. 1 - p. 201

1169. THE PHOSPHODIESTER LINKAGES BETWEEN ADJACENT NUCLEOTIDES IN RNA ARE IN THE 3':5' POSITION BECAUSE OF THE FOLLOWING OBSERVATIONS:

1. Incomplete hydrolysis of RNA with dilute alkali yields a cyclic 3':5' phosphodiester
2. Enzymatic hydrolysis with enzymes specific for the 3' or 5' linkages yields a mixture of 3'- and 5'-phosphates

Ref. 1 - pp. 200-201

1170. IN DNA THE PHOSPHODIESTER LINKAGES MUST BE IN THE 3':5' POSITION BECAUSE:

1. In DNA the pentose has no hydroxyl in the 2' position
2. The 2'- hydroxyl position of the pentose is sterically too hindered for esterification

Ref. 1 - pp. 189-190

1171. IN THE DOUBLE HELIX OF DNA:

1. The 2 strands of DNA are identical
2. The 2 strands of DNA are mirror images

Ref. 1 - p. 193

EACH GROUP OF QUESTIONS BELOW CONSISTS OF 5 LETTERED HEADINGS FOLLOWED BY A LIST OF NUMBERED WORDS OR PHRASES. FOR EACH NUMBERED WORD OR PHRASE SELECT THE ONE LETTERED HEADING THAT IS MOST CLOSELY RELATED TO IT:

A. Ferriprotoporphyrin
B. Etioporphyrin III
C. Chlorophyll a
D. Porphin
E. Ferriprotoporphyrin hydroxide

1172. ___ Basic structure from which hemoglobin, cytochromes and chlorophyll are derived
1173. ___ Derivatives are the most important porphyrins in nature
1174. ___ Hemin
1175. ___ A magnesium porphyrin derivative in which one pyrrole ring is partially reduced
1176. ___ Hematin

Ref. 1 - pp. 166-171; 456; 737-738

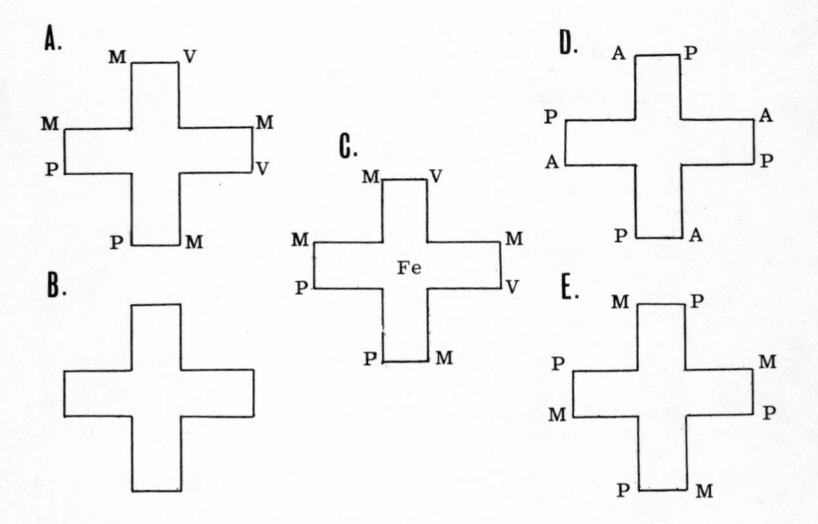

1177. ___ Protoporphyrin Type III (No. IX)
1178. ___ Coproporphyrin I
1179. ___ Heme
1180. ___ Porphin
1181. ___ Uroporphyrin I

Ref. 1 - pp. 166-171; 737-738

CHAPTER IV - BODY FLUIDS
SECTION I - BLOOD

EACH OF THE QUESTIONS OR INCOMPLETE STATEMENTS BELOW IS FOLLOWED BY 5 SUGGESTED ANSWERS OR COMPLETIONS. SELECT THE ONE WHICH IS BEST IN EACH CASE:

1182. THE NATURALLY-OCCURRING PORPHYRINS ARE:
- A. Usually associated with a metallic ion
- B. Usually associated with uncharged metal atoms
- C. Only found in mammalian tissues
- D. Only found in plants
- E. Usually chains of pyrrole rings

Ref. 1 - p. 169

1183. WHICH OF THE FOLLOWING STATEMENTS, REFERRING TO THE PROPERTIES OF PORPHYRINS, IS FALSE?:
- A. They are amphoteric because they contain pyrrole nitrogens and carboxyl groups
- B. They form salts since they contain carboxyl groups
- C. They form esters because they contain carboxyl groups
- D. They are usually colorless compounds
- E. They have an isoelectric point

Ref. 1 - pp. 168-169

1184. CHLOROPHYLL a:
- A. Is identical with ferroprotoporphyrin
- B. Differs from ferroprotoporphyrin only in that it contains Mg
- C. Has a fifth ring which contains no nitrogen
- D. Has free carboxylic acids in the side chain
- E. Has been shown to be identical with pheophytin

Ref. 1 - pp. 456-457

1185. HEMOCHROMOGENS AND PARAHEMATINS ARE:
- A. Porphyrins, the methene bridges of which have been reduced
- B. Formed when hemin is treated with excess alkali
- C. Formed when denatured globin reacts with ferroporphyrin
- D. Mixtures of globin of one species combined with hemin of another species
- E. Formed by reaction of ferro- or ferriporphyrin with 2 moles of a nitrogenous base

Ref. 1 - p. 170

1186. WHICH OF THE FOLLOWING STATEMENTS ABOUT THE STRUCTURE OF HEMOGLOBIN IS UNTRUE?:
- A. Mammalian hemoglobins have a molecular weight of about 65, 000
- B. The molecule of human hemoglobin consists of 4 peptide chains and 4 heme groups
- C. A molecule of adult human hemoglobin contains 2 alpha chains, 2 beta chains and 1 heme group
- D. The conformation of hemoglobin has been elucidated by means of X-ray crystallography
- E. Hemoglobin forms a stable complex with oxygen, leaving the iron in the ferrous state

Ref. 1 - p. 172
Ref. 2 - pp. 596-597

1187. IN SICKLE CELL ANEMIA, THE GLUTAMYL RESIDUE AT POSITION 6 OF THE β-CHAIN IS REPLACED BY A DIFFERENT AMINO ACID RESIDUE. THE AMINO ACID INVOLVED IS:
A. Aspartic acid
B. Threonine
C. Lysine
D. Glutamic acid
E. Valine Ref. 5 - p. 74

1188. BLOOD PLASMA AND BLOOD SERUM DIFFER IN ONE MAJOR RESPECT:
A. Lipid content
B. Concentration of erythrocytes
C. Carbohydrate content
D. Protein content
E. Concentration of Na^+ Ref. 2 - pp. 575, 619

1189. CERTAIN ANTI-COAGULANTS ARE NORMALLY PRESENT IN THE ORGANISM. ONE OF THESE IS:
A. Lipoprotein lipase
B. Dicumarol
C. Hyaluronidase
D. Chonodroitin
E. Heparin Ref. 2 - p. 629

1190. WHICH OF THE FOLLOWING STATEMENTS ABOUT ANTIBODIES IS INCORRECT?:
A. Some of the antibodies in plasma are associated with the β-globulin fraction
B. Human γ-globulins consist of at least 25 to 30 different proteins
C. A typical immunoglobulin consists of 4 subunits
D. The γ-globulins are formed predominantly by the liver
E. Immunoglobulins have a molecular weight of approximately 150,000 Ref. 2 - pp. 583-587

1191. A CERTAIN DEOXY-SUGAR OCCURS CHARACTERISTICALLY IN BLOOD GROUP POLYSACCHARIDES. THIS SUGAR IS:
A. 2-Deoxy-D-ribose
B. 6-Deoxy-L-mannose
C. L-Fucose
D. L-Rhamnose
E. 2-Deoxy-D-glucose Ref. 1 - pp. 33, 736

1192. THE CATABOLISM OF HEMOGLOBIN INVOLVES:
A. The formation of bile pigments
B. Oxidative cleavage of the porphyrin ring
C. The formation of urobilin
D. The formation of stercobilinogen
E. All of these Ref. 1 - pp. 741-746

1193. METABOLIC DEGRADATION OF HEMOGLOBIN TAKES PLACE PRINCIPALLY IN:
A. The reticulo-endothelial system
B. The erythrocyte
C. The liver parenchyma
D. Kidney tubules
E. All of these Ref. 1 - p. 742

1194. BILIRUBIN SECRETED INTO THE INTESTINE IS SUBJECTED TO ENZYMATIC DEGRADATION; THE FINAL PRODUCT BEING:
A. Stercobilin
B. Mesobilirubinogen
C. Biliverdin
D. Bilirubinogen
E. Urobilinogen Ref. 1 - p. 744

1195. WHICH IS NOT A PART OF THE HEMOGLOBIN MOLECULE ?:
A. Histidine
B. Vinyl groups
C. Ferric ion
D. Pyrrole rings
E. Protein Ref. 1 - pp. 171-175,737

1196. THE ISOHYDRIC SHIFT PRODUCES:
A. An increase in the ratio of red cell bicarbonate to plasma bicarbonate
B. The same ratio of bicarbonate to chloride in both red cell and plasma
C. An increase in red cell oxyhemoglobin
D. The principal means for CO_2 transport in the blood
E. None of these Ref. 1 - p. 769

1197. THE OXYGEN DISSOCIATION CURVE IS SHIFTED TO THE RIGHT BY:
A. Decreased CO_2 tension
B. Increased CO_2 tension
C. Increased pH
D. Increased N_2 tension
E. Decreased N_2 tension Ref. 1 - pp. 760-762

1198. __________CAN PREVENT BLOOD CLOTTING:
A. Calcium chloride
B. Potassium cyanide
C. Sodium citrate
D. Protamine sulfate
E. None of these
Ref. 1 - p. 725

1199. THE MATURE ERYTHROCYTE POSSESSES:
A. Glycolytic enzymes
B. Enzymes of tricarboxylic acid cycle
C. Both A and B
D. Hexose monophosphate shunt - but not A or B
E. Hexose monophosphate shunt as well as A and B
Ref. 1 - pp. 739-740

1200. MYOGLOBIN:
A. Exhibits the Bohr effect
B. Has an oxygen dissociation curve which is constant over a wide range of pH
C. Is an auxiliary system for oxygen transport in the blood
D. Contains four hemes per molecule
E. Has none of these properties
Ref. 1 - p. 772

1201. COPROPORPHYRIN III IS:
A. Normally excreted in urine
B. Excreted in lead poisoning
C. An abnormal hemoglobin precursor
D. A breakdown product in normal hemoglobin metabolism
E. Excreted in hepatic porphyria
Ref. 1 - pp. 737; 845

1202. WHICH OF THE FOLLOWING IS NOT AN ANTICOAGULANT?:
A. EDTA
B. Vitamin K
C. Sodium oxalate
D. Heparin
E. Dicumarol
Ref. 1 - pp. 728; 731

1203. MATURE ERYTHROCYTES DO NOT CONTAIN:
A. Glycolytic enzymes
B. Enzymes of the hexose phosphate shunt
C. Pyridine nucleotides
D. ATP
E. Enzymes of the Krebs cycle
Ref. 1 - pp. 739-740

1204. THE NORMAL pH OF BLOOD IS:
A. 6.8
B. 7.1
C. 7.4
D. 7.7
E. 8.0
Ref. 1 - p. 793

1205. THE "CHLORIDE-SHIFT" IMPLIES THAT:
A. For each bicarbonate ion leaving the red cell a potassium ion goes along to preserve electrostatic neutrality
B. Bicarbonate leaves the red cell in the form of H_2CO_3 which is uncharged
C. Bicarbonate leaving the red cell is replaced by chloride
D. For each molecule of CO_2 entering the cell one hydrogen ion leaves the cell
E. Sodium chloride leaves the cell when CO_2 is discharged at the respiratory organ
Ref. 1 - pp. 769; 798

1206. CHRISTMAS FACTOR IS SYNONYMOUS WITH:
A. Proconvertin
B. Platelet accelerator
C. Accelerator globulin
D. Antihemophilic factor
E. Thrombokinase
Ref. 1 - p. 729

1207. WHICH OF THE FOLLOWING AMINO ACIDS IS AN IMPORTANT PRECURSOR OF PROTOPORPHYRIN?:
A. Alanine
B. Proline
C. Glycine
D. Leucine
E. Histidine
Ref. 1 - pp. 584-585; 737

1208. DIRECTLY AND INDIRECTLY HEMOGLOBIN IS RESPONSIBLE FOR THE TRANSPORT OF ________ % OF THE CO_2 CARRIED BY THE BLOOD:
A. 10
B. 50
C. 5
D. 90
E. 100
Ref. 1 - p. 770

1209. THE MATURE ERYTHROCYTE CONTAINS:
A. DNA only
B. RNA only
C. DNA and RNA
D. Little or no DNA or RNA
E. Mostly RNA with a very small amount of DNA
Ref. 1 - pp. 739-740

1210. ERYTHROCYTES OF PERSONS WITH SICKLE CELL ANEMIA CONTAIN:
A. Only hemoglobin A
B. Little hemoglobin S
C. Little hemoglobin A
D. Both hemoglobin A and S
E. Only hemoglobin F
Ref. 1 - pp. 656; 688

1211. HEMOCYANIN:
A. Is a blue copper-containing molecule, concerned with oxygen transport in certain invertebrates
B. Is the compound which results from the reaction of cyanide and hemoglobin
C. Is an oxygen-carrying pigment containing no heme but four times as much iron as hemoglobin
D. Is a pigment released by hemolyzed erythrocytes
E. Contains both Cu and Fe
Ref. 1 - p. 773

1212. CRIGLER-NAJJAR SYNDROME IS A CONGENITAL ABNORMALITY OF THE LIVER CHARACTERIZED BY ACCUMULATION OF BILIRUBIN IN THE BLOOD. THE CAUSE OF THE DISEASE IS THE LACK OF AN ENZYME WHICH CONJUGATES BILIRUBIN. THE ENZYME IS:
A. Uridine diphosphate glucose epimerase
B. Phosphogalactose uridyl transferase
C. Uridine diphosphate glucuronate transferase
D. Phosphoglucose uridyl transferase
E. Uridine diphosphate galactose pyrophosphorylase
Ref. 1 - p. 746

1213. IN HEPATIC PORPHYRIA:
A. Excessive and abnormal formation of heme precursors occurs in liver
B. Excessive and abnormal formation of heme precursors occurs in developing red blood cells of bone marrow
C. Skin photosensitivity occurs
D. Coproporphyrin III is excreted in the urine
E. All of the above apply
Ref. 1 - p. 739

1214. MATURE ERYTHROCYTES:
A. Contain cholesterol largely in the esterified form
B. Contain cholesterol but cannot synthesize it
C. Synthesize cholesterol but lose it by exchange with plasma
D. Cannot synthesize glutathione
E. Contain mitochondria and glutathione
Ref. 1 - pp. 735; 739

1215. THE SYNTHESIS OF ANTIBODY PROTEIN TAKES PLACE:
A. By the general mechanism of protein synthesis
B. On a single type of polyribosome
C. On H chains only
D. On lysosomes
E. On tRNA
Ref. 1 - p. 717

EACH GROUP OF QUESTIONS BELOW CONSISTS OF 5 LETTERED HEADINGS FOLLOWED BY A LIST OF NUMBERED WORDS OR PHRASES. FOR EACH NUMBERED WORD OR PHRASE SELECT THE ONE LETTERED HEADING THAT IS MOST CLOSELY RELATED TO IT:

A. Prothrombin
B. Heparin
C. Proconvertin
D. Fibrinogen
E. Thromboplastin

1216.___ Coagulated by strains of staphylococci, certain snake venoms and thrombin; least soluble of all plasma proteins
1217.___ A glycoprotein containing glucosamine which migrates with $alpha_2$ - globulins in electrophoretic analysis
1218.___ Thrombokinase
1219.___ A relatively stable substance which persists in stored plasma and acts with thromboplastin to hasten prothrombin conversion
1220.___ An acid mucopolysaccharide containing glucosamine, sulfuric acid and glucuronic acid

Ref. 1 - pp. 726-729; 731-733

A. Glucose-6-phosphate dehydrogenase of serum
B. Alkaline phosphatase of serum
C. Acid phosphatase of serum
D. Transaminases of serum
E. Amylase of serum

1221.___ Elevated in prostatic carcinoma
1222.___ Elevated after myocardial infarction
1223.___ Increased in bone disease
1224.___ Elevated in biliary obstruction
1225.___ Elevated in acute pancreatitis
1226.___ Not present in appreciable concentrations

Ref. 1 - p. 723

DEFICIENCIES OF CERTAIN COAGULATION FACTORS CAUSE DISEASE IN MAN. MATCH THE DISORDER OBSERVED WITH THE PERTINENT COAGULATION FACTORS LISTED BELOW:

A. Hemophilia B
B. Willebrand's disease
C. Thrombophilia
D. Thrombocytopenia
E. Parahemophilia

1227.___ Autoprothrombin C
1228.___ Platelet cofactor (F-VIII)
1229.___ Ac-globulin
1230.___ Antithrombin
1231.___ Platelet factor 3

Ref. 2 - pp. 626-627

ANSWER THE FOLLOWING GROUP OF QUESTIONS BY USING THE KEY OUTLINED BELOW:

A. If the item is associated with A only
B. If the item is associated with B only
C. If the item is associated with both A and B
D. If the item is associated with neither A nor B

A. Porphyrins
B. Bile pigments
C. Both
D. Neither

1232.___ Contain a tetrapyrrole chain
1233.___ Contain four pyrrole rings linked by four carbon atoms in a closed-ring system
1234.___ Verdohemochrome
1235.___ Rupture of alpha-methene bridge in hemoglobin yields ______________
1236.___ React(s) with uridine diphosphoglucuronate to form the corresponding diglucuronide
1237.___ Contain(s) Fe^{++}
1238.___ Contain(s) pyrrole rings

Ref. 1 - pp. 742-743

A. Erythrocyte DPNH
B. Erythrocyte TPNH
C. Both
D. Neither

1239.___ Arise(s) from glycolysis
1240.___ Arise(s) from phosphogluconate oxidative pathway
1241.___ Oxidized by methemoglobin reductases
1242.___ Oxidized by cytochrome c
1243.___ Co-factor of glutathione reductase

Ref. 1 - p. 740

A. Biliverdin
B. Bilirubin
C. Both
D. Neither

1244.___ Found in normal human blood
1245.___ Reduced by an enzyme system in intestine
1246.___ Metabolite of hemoglobin
1247.___ Determined in serum by Van den Bergh reaction
1248.___ Determined with Ehrlich's aldehyde reagent
1249.___ Able to bind iron

Ref. 1 - pp. 107; 742-745

A. Hemoglobin
B. Methemoglobin
C. Both
D. Neither

1250.___ Cannot function as oxygen carrier
1251.___ Same as hematin
1252.___ Contain(s) iron in ferric state
1253.___ Contain(s) iron in ferrous state
1254.___ Valence of iron does not change when combining reversibly with oxygen
1255.___ Present in erythrocytes lacking methemoglobin reductase
1256.___ Contains CO

Ref. 1 - pp. 170-172; 177; 740; 758

A. Ferritin
B. Transferrin
C. Both
D. Neither

1257. ___ A compound of a protein and a ferric hydroxide-phosphate
1258. ___ Increase(s) 20 to 50 fold in intestinal mucosa of fasted guinea pigs after iron administration
1259. ___ A $beta_1$-globulin which facilitates iron transport in plasma
1260. ___ Contain(s) iron in the ferric form
1261. ___ Also called ceruloplasmin
1262. ___ Also called hemosiderin
1263. ___ Can bind Cu and Fe

Ref. 1 - pp. 721; 748

A. l-Stercobilin
B. d-Urobilinogen
C. Both
D. Neither

1264. ___ Initial reduction product of bilirubin in the intestine
1265. ___ Reduced to biliverdin in the intestine
1266. ___ Primarily responsible for brown color of feces
1267. ___ Oxidation product of urobilin
1268. ___ Metabolite of hemoglobin

Ref. 1 - pp. 743-744

MATCH THE ELECTROPHORETIC PATTERNS OF SERUM WITH THE CORRECT DIAGNOSIS:

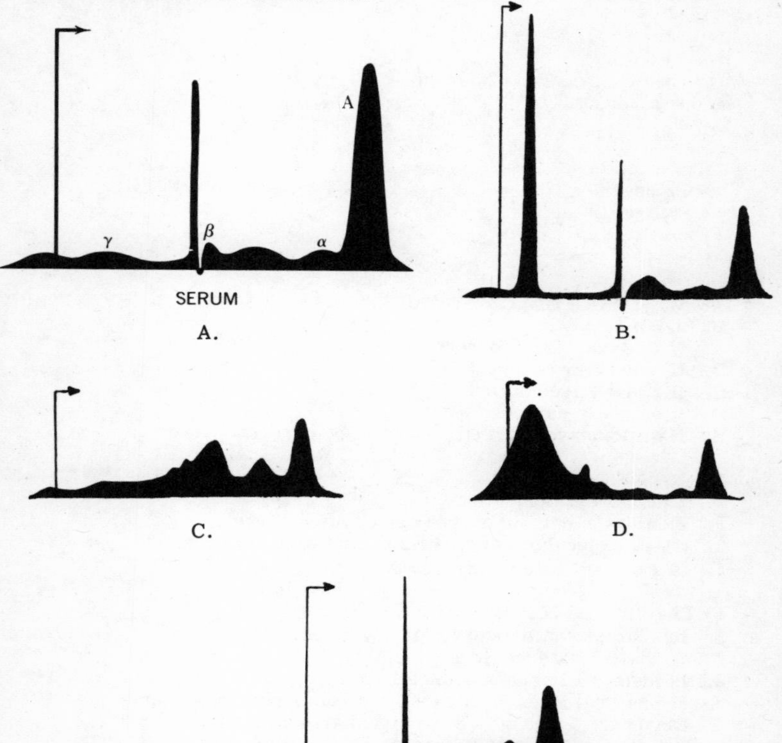

1269. ___ Multiple myeloma

1270. ___ Cirrhosis

1271. ___ Hodgkin's disease

1272. ___ Normal serum

1273. ___ Nephrosis

Ref. 1 - pp. 725; 708

MATCH THE COMPONENTS OF PLASMA LISTED BELOW WITH THEIR CORRECT ELECTROPHORETIC MOBILITIES:

A. α_1
B. α_2
C. β_1
D. β_2
E. γ

1274. ___ Transcortin
1275. ___ Fibrinogen
1276. ___ Ceruloplasmin
1277. ___ Cryoglobulin
1278. ___ Glycoprotein Ref. 2 - p. 581

ANSWER THE FOLLOWING QUESTIONS BY USING THE KEY OUTLINED BELOW:
A. If 1, 2 and 3 are correct
B. If 1 and 3 are correct
C. If 2 and 4 are correct
D. If only 4 is correct
E. If all four are correct

1279. MYOGLOBIN:
1. Exhibits the Bohr effect
2. Exhibits a sigmoid oxygen dissociation curve
3. Has a molecular weight 4 times that of hemoglobin
4. Is a storage site for oxygen in muscle
Ref. 1 - pp. 243, 761, 772

1280. FETAL HEMOGLOBIN:
1. Has a dissociation curve for any given value of CO_2 and O_2 identical to that of maternal hemoglobin
2. Is identical immunochemically to maternal hemoglobin
3. Is identical physicochemically to maternal hemoglobin
4. Disappears gradually from circulation after birth
Ref. 1 - pp. 771-772

1281.
1. Hemoglobin is a buffer
2. The buffering action of hemoglobin is primarily due to -SH groups of histidine residues
3. Oxyhemoglobin is a stronger acid than reduced hemoglobin
4. At physiological pH the ratio of HCO_3^-/H_2CO_3 is 1/20
Ref. 1 - pp. 763; 767

1282.
1. An adult man will produce an amount of H_2CO_3 (from CO_2) each day equivalent to 2 liters of concentrated HCl
2. The major portion of CO_2 transported in the blood is in the form of bicarbonate
3. Hydration of CO_2 in erythrocytes is catalyzed by the enzyme carbonic anhydrase
4. There are three times as many carbamino groups on oxyhemoglobin as on hemoglobin
Ref. 1 - pp. 758;766-768

1283.
1. The affinity of hemoglobin for oxygen is 200 times that for carbon monoxide
2. Molecules of hemoglobin containing carbon monoxide as well as oxygen give up their oxygen much more easily than does oxyhemoglobin
3. Erythrocytes from persons with hereditary hemolytic anemia, induced by aromatic compounds such as primaquine, have excessively high glucose-6-phosphate dehydrogenase levels
4. Sickling of erythrocytes occurs when oxyhemoglobin S is reduced to hemoglobin S, thereby causing excessive precipitation of hemoglobin S within the erythrocyte

Ref. 1 - pp. 750-753

1284.
1. Leukocytes have enzymes of glycolysis but no cytochromes or Krebs cycle enzymes
2. Leukocytes of patients with Von Gierke's disease have abnormally high glycogen content
3. Injection of adrenal cortical steroids causes an increase in the number of circulating lymphocytes
4. Leukocytes of patients with acute leukemia exhibit high dihydrofolate reductase activity

Ref. 1 - pp. 753-754

1285. CONCERNING THE PLASMA PROTEINS:
1. When protein intake is restricted the albumin/globulin ratio increases
2. Albumin is of greater importance in osmotic regulation than globulins, because of its lower molecular weight
3. The alpha-globulins are glycoproteins which are synthesized in extrahepatic tissues
4. Most of the serum copper is in the form of ceruloplasmin

Ref. 1 - pp. 707; 722-724

1286.
1. In Wilson's disease there is a marked increase in plasma ceruloplasmin levels and in the copper concentration of liver and brain
2. In many infectious diseases the fibrinogen level of the blood increases and alpha-globulin levels decrease
3. In acanthocytosis the erythrocytes have spiny projections due to the absence of serum alpha-lipoprotein
4. Bence-Jones proteins are found in the urine of many multiple myeloma patients. They are similar to human serum gamma-globulin

Ref. 1 - pp. 717; 722-724

1287.
1. When protein intake is restricted the albumin/globulin ratio increases
2. Approximately 50% of total NPN of serum is due to urea and 25% to free amino acids
3. Plasma uric acid is significantly elevated in gout and after the ingestion of purine-rich foods
4. Plasma enzymes (except those involved in coagulation) probably perform no important metabolic function

Ref. 1 - pp. 706; 723

1288. 1. The interior of the erythrocyte is the most concentrated protein solution in the body
2. The adult mammalian erythrocyte has a very active respiratory metabolism
3. The adult erythrocyte contains a considerable quantity of diphosphoglycerate
4. The membrane of the erythrocyte is composed entirely of protein and carbohydrate
Ref. 1 - pp. 734-735; 768

1289. 1. Pronormoblasts contain nuclei and nucleoli and RNA
2. The reticulocyte contains small amounts of RNA
3. The adult erythrocyte contains no detectable RNA or DNA
4. Heme and globin synthesis begin only in the adult erythrocyte
Ref. 1 - p. 735

1290. THE MATURE ERYTHROCYTE CONTAINS:
1. Lecithin
2. Cerebrosides
3. Cholesterol
4. ATP
Ref. 1 - pp. 739-740

1291. EARLY INTERMEDIATES IN THE SYNTHESIS OF HEMOGLOBIN ARE:
1. Glycine
2. Delta-aminolevulinic acid
3. Succinyl CoA
4. Gamma-aminobutyric acid
Ref. 1 - p. 586

1292. THE MATURE ERYTHROCYTE CONTAINS:
1. Cytochromes
2. Enzymes of tricarboxylic acid cycle
3. Succinic dehydrogenase
4. ATPase
Ref. 1 - pp. 739-740

1293. 1. About 0.85% of the red cell mass turns over per day
2. Protein within erythrocytes is in a state of constant turnover
3. The life span of the red cell is about 126 days
4. The ATP content of the erythrocyte is unchanged during storage of blood
Ref. 1 - pp. 740-741

1294. 1. Hemoglobins A and S differ in amino acid composition
2. Hemoblobin S has a neutral valyl residue in place of negatively charged glutamyl residue in hemoglobin A
3. Reduced hemoglobin S is less soluble than hemoglobin A
4. Hemoglobin S has the same isoelectric point as hemoglobin A
Ref. 1 - pp. 172; 750-751

ANSWER THE FOLLOWING QUESTIONS BY USING THE KEY OUTLINED BELOW:

A. If A is greater than B
B. If B is greater than A
C. If A and B are equal or nearly equal

1295. A. Amount of CO_2 in erythrocyte transported as HCO_3^-
B. Amount of CO_2 in erythrocyte transported as carbamino-CO_2
Ref. 1 - pp. 766-767

1296. A. Affinity of hemoglobin for oxygen
B. Affinity of myoglobin for oxygen
Ref. 1 - p. 772

1297. A. Affinity of hemoglobin for CO
B. Affinity of hemoglobin for O_2 Ref. 1 - p. 772

1298. A. Affinity of fetal hemoglobin for O_2
B. Affinity of maternal hemoglobin for O_2
Ref. 1 - pp. 771-772

1299. A. pH of plasma
B. pH of interior of erythrocyte
Ref. 1 - p. 769

1300. A. Arterial HCO_3^- concentration in plasma
B. Arterial HCO_3^- concentration in erythrocyte
Ref. 1 - p. 768

1301. A. Concentration of protein in erythrocytes
B. Concentration of protein in plasma Ref. 1 - p. 768

1302. A. Hematocrit of arterial blood
B. Hematocrit of venous blood Ref. 1 - p. 770

1303. A. Number of normal α-chains in Hb A
B. Number of normal α-chains in Hb S
Ref. 1 - pp. 749-751

1304. A. Number of normal β-chains in Hb S
B. Number of normal α-chains in Hb S
Ref. 1 - pp. 749-751

1305. A. Length of β-chain in hemoglobin
B. Length of δ-chain in hemoglobin Ref. 1 - pp. 749-751

1306. A. Solubility of hemoglobin in normals
B. Solubility of hemoglobin in individuals with sickle cell anemia
Ref. 1 - pp. 749-751

1307. A. Affinity of hemoglobin for CO_2
B. Affinity of hemoglobin for CO Ref. 1 - p. 751

1308. A. Number of gamma-chains in fetal hemoglobin
B. Number of gamma-chains in adult hemoglobin
Ref. 2 - p. 597

MATCH THE FOLLOWING:

WHAT ARE THE APPROXIMATE RANGES OF CONCENTRATIONS OF THE FOLLOWING INORGANIC CONSTITUTENTS OF HUMAN BLOOD PLASMA?:

A. 24-30 meq/liter
B. .05-.18 mg/100 ml
C. 100-110 meq/liter
D. .006-.008 mg/100 ml
E. 132-150 meq/liter

1309. ___ Chloride
1310. ___ Iodine (protein bound)
1311. ___ Bicarbonate
1312. ___ Sodium
1313. ___ Iron

Ref. 1 - p. 707

EACH OF THE QUESTIONS OR INCOMPLETE STATEMENTS BELOW IS FOLLOWED BY 5 SUGGESTED ANSWERS OR COMPLETIONS. SELECT THE ONE WHICH IS BEST IN EACH CASE:

1314. DEHYDRATION CAN OCCUR AS A RESULT OF:
A. Diabetes
B. Chronic glomerulonephritis
C. Addison's disease
D. Diarrhea
E. All of these
Ref. 1 - pp. 506;788-789;845; 958

1315. THE LOWEST pH OF EXTRACELLULAR FLUIDS IS FOUND IN:
A. Plasma
B. Tears
C. Parietal gastric juice
D. Gastric mucus
E. Mixed gastric secretions
Ref. 1 - p. 812

1316. THE MOST IMPORTANT BUFFER IN THE EXTRACELLULAR FLUID IS:
A. Phosphate
B. Protein
C. Acetate
D. Chloride
E. $HCO_3^- - H_2CO_3$
Ref. 1 - p. 790

1317. METABOLIC ACIDOSIS MAY BE BROUGHT ABOUT BY:
A. Loss of CO_2 by increased ventilation
B. Retention of CO_2 by respiratory obstruction
C. Persistent vomiting
D. Absorption of excessive amounts of $NaHCO_3$
E. Presence of acetoacetic acid in extracellular fluid
Ref. 1 - pp. 798-800

1318. ALL OF THE FOLLOWING ARE COSTITUENTS OF NORMAL HUMAN URINE, EXCEPT:
A. Indican
B. Bence Jones protein
C. Uric acid
D. Creatinine
E. Sulfate
Ref. 2 - pp. 727, 740

1319. WHICH OF THE FOLLOWING IS LIKELY TO PRODUCE METABOLIC ACIDOSIS?:
A. High altitude
B. Excessive vomiting
C. Starvation
D. Toxic doses of morphine
E. Ingestion of excessive amounts of sodium bicarbonate
Ref. 2 - pp. 662-663

THE FOLLOWING DIAGRAM DESCRIBES THE EFFECT OF CO_2 TENSION UPON THE DISSOCIATION OF OXYHEMOGLOBIN.

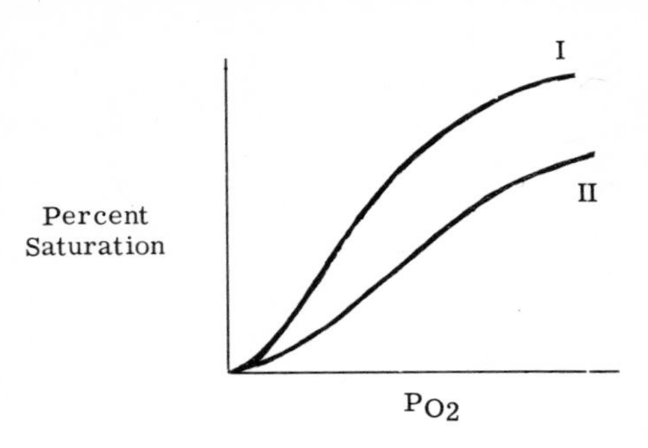

1320. IT CAN BE DEDUCED FROM THIS DIAGRAM THAT:
A. The formation of oxyhemoglobin is not influenced by pH
B. The formation of oxyhemoglobin is influenced by the partial pressure of CO_2
C. The formation of oxyhemoglobin is influenced by salt concentration
D. Curve II describes a lower P_{CO_2} than curve I
E. Curves I and II represent lines of equal P_{CO_2}

Ref. 1 - pp. 760-762

1321. THE BOHR EFFECT DESCRIBES:
A. The effect of P_{CO_2} on the dissociation of oxyhemoglobin
B. The effect of Zn^{++} on the activity of carbonic anhydrase
C. The noxious effect of increased P_{O_2} upon the nervous system
D. The buffering action of imidazole groups
E. All of the above Ref. 1 - p. 763

1322. ADDISON'S DISEASE:
A. Is not associated with an electrolyte disturbance
B. Is associated with hypernatremia
C. Is associated with hyperpotassemia
D. Is associated with hypopotassemia
E. Is associated with hyponatremia Ref. 1 - p. 802

1323. A SOLUTION CONTAINS 0.53 MG. OF Na_2CO_3 PER ML. WHAT IS THE CONCENTRATION OF Na^+ IN MEQ/LITER (H = 1; Na = 23; C = 12; O = 16)?:
A. 0.1
B. 1
C. 10
D. 100
E. 1000 Ref. 1 - p. 758

1324. ELEVATION OF BLOOD PRESSURE WOULD MOST LIKELY BE PRODUCED BY:
A. Ketone bodies
B. Urea
C. Hyaluronidase
D. Renin
E. Purine
Ref. 1 - p. 839

1325. IN MAN, THE PRINCIPAL NITROGENOUS EXCRETION PRODUCT(S) IS (ARE):
A. Amino acids
B. Ammonia
C. Uric acid
D. Purines and pyrimidines
E. None of these
Ref. 1 - pp. 841-842

1326. THE PRIMARY SOURCE OF THE NITROGEN OF URINARY UREA IS:
A. Ammonia derived from purine degradation
B. Nitrogen derived from purines and pyrimidines
C. Ornithine
D. Citrulline
E. Ammonia derived from amino acid deamination reactions and transamination reactions
Ref. 1 - pp. 560-561

EACH GROUP OF QUESTIONS BELOW CONSISTS OF 5 LETTERED HEADINGS FOLLOWED BY A LIST OF NUMBERED WORDS OR PHRASES. FOR EACH NUMBERED WORD OR PHRASE SELECT THE ONE LETTERED HEADING THAT IS MOST CLOSELY ASSOCIATED TO IT:

A. Na^+
B. K^+, Mg^{++}
C. Cl^-, HCO_3^-
D. $PO_4^{\equiv}$, $SO_4^=$, protein
E. Cl^-, $SO_4^=$

1327. ___ Chief intracellular cations
1328. ___ Chief extracellular cations
1329. ___ Chief extracellular anions
1330. ___ Chief intracellular anions
1331. ___ Chief anions in sea water

Ref. 1 - p. 779

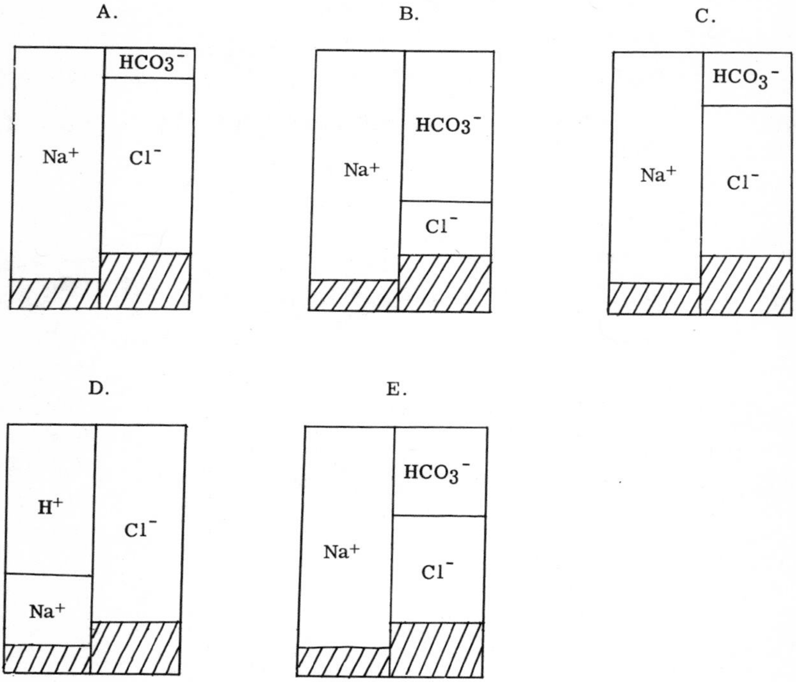

A.

Na^+ | HCO_3^-, Cl^-

B.

Na^+ | HCO_3^-, Cl^-

C.

Na^+ | HCO_3^-, Cl^-

D.

H^+, Na^+ | Cl^-

E.

Na^+ | HCO_3^-, Cl^-

THE BAR GRAPHS ABOVE REPRESENT ELECTROLYTE COMPOSITION OF:

1332. ___ Serum in metabolic acidosis
1333. ___ Normal serum
1334. ___ Gastric juice
1335. ___ Serum in metabolic alkalosis
1336. ___ Pancreatic juice

Ref. 1 - pp. 788; 799

A. $H_2PO_4^-$
B. Alveolar CO_2 tension
C. H_2CO_3
D. $[HCO_3^-]$
E. NH_4^+

1337. ___ Blood H_2CO_3 concentration is determined by ________
1338. ___ The kidney regulates pH of plasma by regulating ________
1339. ___ Respiratory compensation for variations in plasma pH is exerted through control of plasma ________
1340. ___ ________ constitutes most of the acid conventionally measured as the "titratable acidity" of urine
1341. ___ A renal mechanism for restoring normal extracellular pH in acidotic states is the formation and excretion of ________

Ref. 1 - pp. 790-795; 797

A. Isotonic contraction
B. Hypertonic expansion
C. Isotonic expansion
D. Hypotonic expansion
E. Hypertonic contraction

1342. ___ Accumulation of extracellular water without an equivalent amount of salt
1343. ___ Ingestion of sea water causes ________ of extracellular fluid
1344. ___ Pulmonary edema or palpable edema of the extremities may result from ________ of the extracellular fluid
1345. ___ Diarrhea leads to ________ of the extracellular fluid
1346. ___ Diabetes insipidus may lead to ________ of the extracellular fluid

Ref. 1 - pp. 786-789

A. Na^+
B. Alkaline
C. Acidic
D. Acidosis
E. Alkalosis

1347. ___ Formation and excretion of ammonium ion by the kidney is a mechanism for conservation of ________
1348. ___ Failure of ammonium-forming mechanism in lower nephron nephrosis and Fanconi's syndrome may cause ________ and dehydration
1349. ___ In ________ caused by an increased extracellular $[HCO_3^-]$, Na^+ enters cells in exchange for H^+ and K^+. CO_2 resulting from reaction of the H^+ with HCO_3^- is expired
1350. ___ The ash of most foodstuffs is ________
1351. ___ A diet consisting largely of fruits and vegetables results in excretion of ________ urine

Ref. 1 - pp. 795-797

A. 5%
B. 15%
C. 20%
D. 50%
E. 70%

1352. ___ Total amount of fluid as % of body weight
1353. ___ Interstitial fluid, % of body weight
1354. ___ Blood plasma, % of body weight
1355. ___ Extracellular fluid, % of body weight Ref. 1 - pp. 777, 828
1356. ___ Intracellular fluid, % of body weight Ref. 2 - pp. 440-442

CERTAIN HEREDITARY METABOLIC DISORDERS CAN BE DIAGNOSED BY DETECTING CHARACTERISTIC URINARY METABOLITES. MATCH THE INBORN ERRORS OF METABOLISM WITH THE URINARY METABOLITES CHARACTERISTIC OF THE DISORDER:

A. Uric acid
B. Bence Jones protein
C. Phenylpyruvic acid
D. Glucose
E. δ-Aminolevulinic acid

1357. ___ Multiple myeloma
1358. ___ Phenylketonuria
1359. ___ Diabetes mellitus
1360. ___ Gout
1361. ___ Acute intermittent porphyria Ref. 2 - p. 747

ANSWER THE FOLLOWING GROUP OF QUESTIONS BY USING THE KEY OUTLINED BELOW:
A. If the item is associated with A only
B. If the item is associated with B only
C. If the item is associated with both A and B
D. If the item is associated with neither A nor B

A. Respiratory acidosis
B. Metabolic acidosis
C. Both
D. Neither

1362. ___ Plasma $[HCO_3^-]$ increases
1363. ___ Plasma $[H_2CO_3]$ decreases
1364. ___ Urine pH decreases
1365. ___ Urine pH increases
1366. ___ Plasma $[HCO_3^-]$ decreases Ref. 1 - pp. 796-800

A. Respiratory alkalosis
B. Metabolic alkalosis
C. Both
D. Neither

1367. ___ Urine pH increases
1368. ___ Plasma $[HCO_3^-]$ increases
1369. ___ Plasma $[H_2CO_3]$ decreases
1370. ___ Plasma $[H_2CO_3^-]$ increases
1371. ___ Urine pH decreases Ref. 1 - pp. 796-800

ANSWER THE FOLLOWING QUESTIONS BY USING THE KEY OUTLINED BELOW:
A. If A is greater than B
B. If B is greater than A
C. If A and B are equal or nearly equal

1372. A. $[HPO_4^{=}]/[H_2PO_4^-]$ in extracellular fluid at pH 7.4
B. $[HPO_4^{=}]/[H_2PO_4^-]$ in urine at pH 5.4
Ref. 1 - pp. 793-794

1373. A. Osmotic pressure within cells
B. Osmotic pressure of fluid surrounding cells
Ref. 1 - pp. 784-785

1374. A. Potassium concentration of plasma
B. Potassium concentration of saliva Ref. 1 - pp. 803,815

1375. A. Bicarbonate concentration of pancreatic secretion
B. Bicarbonate concentration of plasma Ref. 1 - p. 818

1376. A. Urinary phosphate concentration
B. Urinary chloride concentration Ref. 1 - pp. 841-842

1377. A. Urinary phosphate concentration
B. Serum phosphate concentration Ref. 1 - pp. 707,841

1378. A. Bicarbonate concentration of gastric juice
B. Bicarbonate concentration of serum Ref. 1 - pp. 707,812

1379. A. Calcium concentration of bile
B. Calcium concentration of plasma Ref. 1 - p. 812

1380. A. Daily secretion of sweat, in ml
B. Daily secretion of pancreatic juice, in ml
Ref. 1 - pp. 813,818

1381. A. Amount of lysozyme in pancreatic juice
B. Amount of lysozyme in tears Ref. 1 - pp. 814,818

ANSWER THE FOLLOWING QUESTIONS BY USING THE KEY OUTLINED BELOW:
A. If 1, 2 and 3 are correct
B. If 1 and 3 are correct
C. If 2 and 4 are correct
D. If only 4 is correct
E. If all four are correct

1382.
1. Glucose levels in cerebrospinal fluid reflect changes in plasma glucose levels
2. NPN constituents are appreciably higher in cerebrospinal fluid than in plasma
3. In cystic fibrosis sweat and tears are characteristically rich in NaCl
4. Tears do not contain protein

Ref. 1 - pp. 811-813

1383.
1. Maintenance of Na^+: K^+ concentration differences across red cell membrane is dependent upon glycolysis
2. The composition of extracellular fluid is similar to that of the seas during the pre-Cambrian era
3. After several days on a diet devoid of Na^+ and Cl^- the urine becomes virtually free of these ions
4. Alterations in osmotic pressure of intracellular fluid have no important physiological effects

Ref. 1 - pp. 779-781

1384. CONCERNING THE COMPOSITION OF URINE:
1. Normally, chloride is the chief anion of urine
2. Virtually all phosphorus in urine is present in combination with organic compounds
3. Since Na^+ and K^+ are the major cations of the diet, they are also the major cations of human urine
4. Hippuric acid is normally not present in human urine

Ref. 1 - pp. 841-843

1385.
1. Na^+ is the chief extracellular cation
2. Mg^{++} is the chief intracellular cation
3. Chloride and bicarbonate are the chief extracellular anions
4. Bicarbonate is the chief intracellular anion

Ref. 1 - p. 779

1386.
1. The aqueous humor has a low protein content
2. Only water and nonelectrolytes may be exchanged between the iris and the aqueous humor
3. Vitreous humor contains a hyaluronic acid gel originally secreted by the retina
4. The potassium and sodium contents of aqueous humor are identical to that of plasma Ref. 1 - pp. 810-811

1387.
1. "Nutritional" edema is always due to hypoalbuminemia
2. Interstitial fluid has slightly lower concentrations of anions and lower concentrations of cations than plasma, in accord with Gibbs-Donnan formulation
3. The protein composition of lymph is the same as that of plasma
4. The synovial fluid contains a mucoprotein, the polysaccharide component of which is hyaluronic acid

Ref. 1 - pp. 808-809

1388. ABNORMAL CONSTITUENTS OF URINE:
1. Normal urine gives a positive test for protein with the usual clinical methods
2. An appreciable proportion of patients with severe hyperthyroidism have glucosuria
3. Following the ingestion of avocados, fructose, a reducing sugar, may appear in the urine
4. Patients with idiopathic pentosuria show positive tests for reducing sugar Ref. 1 - pp. 844-845

CHAPTER V - NUTRITION
SECTION I - DIET, ABSORPTION AND DIGESTION

EACH OF THE QUESTIONS OR INCOMPLETE STATEMENTS BELOW IS FOLLOWED BY 5 SUGGESTED ANSWERS OR COMPLETIONS. SELECT THE ONE WHICH IS BEST IN EACH CASE:

1389. THE OXIDATION OF WHICH SUBSTANCE YIELDS THE LARGEST NUMBER OF CALORIES PER GRAM?:
A. Lipids
B. Plant protein
C. Glucose
D. Animal protein
E. Glycogen
Ref. 1 - p. 293

1390. A NEGATIVE NITROGEN BALANCE IS OBSERVED:
A. During normal pregnancy
B. During normal child growth
C. During convalescence
D. In malnutrition
E. In none of these
Ref. 1 - p. 555

1391. THE INGESTION OF WHICH FOODSTUFF RESULTS IN THE GREATEST SPECIFIC DYNAMIC EFFECT?:
A. Fat
B. Carbohydrate
C. Protein
D. Vitamins
E. Water
Ref. 1 - p. 299

1392. AMONG THE ELEMENTS NEEDED BY THE BODY ARE:
A. Iodine
B. Copper
C. Manganese
D. Chlorine
E. All of these
Ref. 1 - p. 1014

1393. THE BIOLOGICAL VALUE OF A GIVEN PROTEIN DEPENDS UPON:
A. The digestibility alone
B. Digestibility and amino acid composition
C. Amino acid composition alone
D. Amino acid composition and specific dynamic effect
E. Digestibility and leucine content
Ref. 1 - pp. 1006-1007

1394. ABSORPTIVE LIPEMIA IS DUE TO INGESTION OF:
A. Sugars
B. Proteins
C. Starches
D. Fats
E. Calcium
Ref. 1 - pp. 475-476

1395. PROTEIN DIGESTION IN THE STOMACH INVOLVES THE ACTION OF THE ENZYME:
A. Ptyalin
B. Lipase
C. Pepsin
D. Mucin
E. Enterokinase
Ref. 1 - p. 532

1396. THE RECOMMENDED DAILY DIETARY ALLOWANCE OF PROTEIN FOR AN AVERAGE 70 Kg MALE IS CURRENTLY SET AT:
A. 6.5 g
B. 0.09 g per Kg body weight
C. 110 g
D. 13 g
E. 65 g
Ref. 2 - pp. 841-844

SECTION I - DIET, ABSORPTION AND DIGESTION

1397. THE MAJOR PATHWAY OF CALCIUM EXCRETION UNDER NORMAL CONDITIONS IS:
A. Intestinal tract
B. Kidneys
C. Both A and B
D. Sweat
E. All of these
Ref. 1 - p. 889

1398. TETANY DUE TO LOW SERUM CALCIUM LEVELS RESULTS FROM REMOVAL OF WHICH GLAND?:
A. Thyroid
B. Adrenal cortex
C. Pituitary
D. Parathyroid
E. Thyroid
Ref. 1 - pp. 888; 934

1399. WHICH OF THE FOLLOWING STATEMENTS CONCERNING CALCIUM METABOLISM IS MOST CORRECT?:
A. Non-diffusible plasma calcium is primarily calcium phosphate
B. Normal human plasma contains 1 to 4 meq. Ca^{++} per liter
C. Plasma low in protein is also low in calcium
D. In normal individuals 50% of the calcium filtered through the glomerulus is excreted
E. Increasing plasma pH increases the concentration of ionized calcium
Ref. 1 - p. 888

1400. THE ABSORPTION OF IRON IS UNIQUE IN THAT:
A. It is regulated by the level of transferrin in the intestinal mucosa
B. It requires a specific protein, hemosiderin
C. It is completely absorbed
D. It forms a complex with the dietary proteins before it can be absorbed
E. Absorption must be exactly adjusted to need since there is no excretory pathway of excess iron
Ref. 1 - p. 749

1401. AN IMPORTANT ETIOLOGIC FACTOR IN KWASHIORKOR IS:
A. Steatorrhea
B. Atrophy of acinar tissue of pancreas
C. Anemia
D. Dietary protein deficiency
E. Excess dietary ethionine
Ref. 1 - p. 1009

1402. A SENSITIVE ENZYME TEST FOR DETERMINING THE BIOLOGICAL VALUE OF PROTEINS IS MEASUREMENT OF:
A. Liver xanthine oxidase
B. Pepsin activity of gastric juice
C. Serum transaminase
D. Pancreatic amylase
E. Serum lipoprotein lipase
Ref. 1 - p. 1010

1403. WHICH OF THE FOLLOWING STATEMENTS IS MOST NEARLY CORRECT? DURING STARVATION:
A. Liver AMP kinase activity is increased
B. Liver glucokinase activity is increased
C. Activity of pancreatic enzymes is increased
D. The activity of hepatic CoA-carnitine transferase is decreased
E. The activity of hepatic glutamate-alanine transaminase is decreased
Ref. 5 - p. 549

1404. EXPERIMENTAL ATHEROSCLEROSIS IN MONKEYS CAN BE PRODUCED BY:
A. Riboflavin deficiency
B. Pyridoxine deficiency
C. Feeding pyridoxine plus linoleic acid
D. Feeding riboflavin plus arachidonic acid
E. All of these Ref. 1 - p. 1013

EACH GROUP OF QUESTIONS BELOW CONSISTS OF 5 LETTERED HEADINGS FOLLOWED BY A LIST OF NUMBERED WORDS OR PHRASES. FOR EACH NUMBERED WORD OR PHRASE SELECT THE ONE LETTERED HEADING THAT IS MOST CLOSELY RELATED TO IT:

A. 2-monoglyceride
B. Oligo-1, 6-glucosidase
C. Pancreatic lipase
D. Amylase
E. Bile salt

1405. ___ An enzyme present in intestinal mucosa that makes possible complete digestion of ingested amylopectins and glycogen to free glucose
1406. ___ Attacks alpha-1, 4-glucosidic bonds
1407. ___ Hydrolyzes triglycerides and esters of long chain fatty acids in intestine
1408. ___ Increases activity of pancreatic lipase by virtue of detergent action
1409. ___ Appears in intestine during fat digestion as a result of digestion of triglycerides Ref. 1 - pp. 387, 471

RELATE THE MANIFESTATIONS OF SINGLE AMINO ACID DEFICIENCIES IN YOUNG RATS TO THE SPECIFIC AMINO ACIDS INVOLVED:

A. Valine
B. Cystine
C. Leucine
D. Threonine
E. Methionine

1410. ___ Hypoproteinemia
1411. ___ Fatty liver, cirrhosis
1412. ___ Edema
1413. ___ Locomotor dysfunction
1414. ___ Acute hepatic necrosis Ref. 1 - p. 1008

MATCH THE CORRECT MODULATING COMPOUND WITH THE ENZYME WHICH IT INHIBITS:

A. ADP
B. ATP
C. Alanine
D. Acetyl-CoA
E. AMP

1415. ___ Liver pyruvate kinase
1416. ___ Citrate synthase
1417. ___ Fructose diphosphatase
1418. ___ Pyruvate carboxylase
1419. ___ Pyruvate dehydrogenase Ref. 5 - p. 544

MATCH THE CORRECT MODULATING COMPOUND LISTED BELOW WITH THE ENZYME WHICH IT ACTIVATES:

A. Glucose-6-phosphate
B. ADP
C. Fructose-1, 6-diphosphate
D. Acetyl CoA
E. Citrate

1420. ___ Pyruvate carboxylase
1421. ___ Glutamate dehydrogenase
1422. ___ Glycogen synthase D
1423. ___ Phosphofructokinase
1424. ___ Acetyl-CoA carboxylase Ref. 5 - p. 544

A. Cobalt
B. Copper
C. Molybdenum
D. Manganese
E. Zinc

1425. ___ Deficiency causes perosis in chicks
1426. ___ Deficiency is observed in infants purely on milk diet, resulting in microcytic, normochromic anemia
1427. ___ Parenteral administration causes polycythemia
1428. ___ A constituent of alcohol and lactic dehydrogenases
1429. ___ A constituent of xanthine oxidase; deficiency results in decrease of this enzyme in rat liver Ref. 1 - pp. 1014-1017

A. Phospholipase A
B. Bile acid
C. Cholesterol
D. Chylomicron
E. Beta-lipoprotein

1430. ___ Its action produces lysolecithin
1431. ___ Part of intestinal ________ is absorbed in unesterified form in lymph
1432. ___ Consists largely of triglyceride in combination with $alpha_2$-globulin
1433. ___ Cause of absorptive lipemia
1434. ___ Absorbed via portal vein Ref. 1 - pp. 472-475

ARRANGE THE FOLLOWING OCCUPATIONS IN ORDER OF APPROXIMATE CALORIC EXPENDITURE:

A. Highest
B. 2nd highest
C. 3rd highest
D. 4th highest
E. Lowest

1435. ___ Carpenter
1436. ___ Tailor
1437. ___ Shoemaker
1438. ___ Lumberman
1439. ___ Stonemason Ref. 2 - p. 534

MATCH THE CORRECT METABOLIC RATE WITH THE CONDITIONS LISTED BELOW:

A. Basal metabolic rate above normal
B. Basal metabolic rate below normal
C. Basal metabolic rate normal or approximately normal

1440. ___ Addison's disease
1441. ___ Fever
1442. ___ Pregnancy
1443. ___ Cushing's syndrome
1444. ___ Hypothyroidism Ref. 2 - pp. 528-529

ARRANGE THE FOLLOWING ACTIVITIES IN ORDER OF APPROXIMATE ENERGY EXPENDITURE OF A 70 Kg MAN:

A. 65 cal/hr
B. 115 cal/hr
C. 200 cal/hr
D. 570 cal/hr
E. 1100 cal/hr

1445. ___ Walking up stairs
1446. ___ Walking slowly
1447. ___ Sleeping
1448. ___ Standing
1449. ___ Running Ref. 2 - p. 533

ANSWER THE FOLLOWING GROUP OF QUESTIONS BY USING THE KEY OUTLINED BELOW:
A. If the item is associated with A only
B. If the item is associated with B only
C. If the item is associated with both A and B
D. If the item is associated with neither A nor B

A. Chymotrypsin
B. Trypsin
C. Both
D. Neither

1450. ___ pH optimum 7-9
1451. ___ Exist(s) as zymogen in tissues
1452. ___ Clot(s) both milk and blood
1453. ___ Hydrolyze(s) peptide bonds involving tyrosine
1454. ___ Strictly an exopeptidase Ref. 1 - pp. 219, 531-534

A. Pentoses
B. Hexoses
C. Both
D. Neither

1455. ___ Absorption inhibited by phlorizin
1456. ___ Diffuse most rapidly
1457. ___ Absorbed from intestinal lumen most rapidly
1458. ___ Active transport across intestinal mucosa
1459. ___ Absorption inhibited by iodoacetate Ref. 1 - pp. 388-389, 837
Ref. 2 - pp. 184-185

A. Pancreatic juice
B. Bile
C. Both
D. Neither

1460. ___ Contain(s) cholesterol
1461. ___ Secretion controlled by hormones of intestinal mucosa
1462. ___ High concentration of digestive enzymes
1463. ___ Low concentration of digestive enzymes
1464. ___ Contain(s) alkaline phosphatase

Ref. 1 - pp. 818-821

A. Carboxypeptidase A
B. Aminopeptidase
C. Both
D. Neither

1465. ___ Present in pancreatic secretion
1466. ___ Contain(s) Zn
1467. ___ Present in intestinal mucosa
1468. ___ Exopeptidase(s)
1469. ___ Endopeptidase(s)

Ref. 1 - pp. 144;219;818
Ref. 2 - pp. 342-344

ANSWER THE FOLLOWING QUESTIONS BY USING THE KEY OUTLINED BELOW:
A. If A is greater than B
B. If B is greater than A
C. If A and B are equal or nearly equal

1470. A. Ability of corn oil in diet to maintain low concentration of serum lipids
B. Ability of lard in diet to maintain low concentration of serum lipids

Ref. 1 - pp. 1012-1013

1471. A. Biological value of plant proteins
B. Biological value of animal proteins

Ref. 1 - pp. 1006-1007

1472. A. Biological value of casein
B. Biological value of lactalbumin

Ref. 1 - p. 1007

1473. A. Heat production resulting from utilization of one liter of oxygen during combustion of fat
B. Heat production resulting from utilization of one liter of oxygen during combustion of carbohydrates

Ref. 1 - pp. 294-295

1474. A. Normal urinary excretion of Ca
B. Normal fecal excretion of Ca

Ref. 1 - p. 889

1475. A. Percentage of body Ca present in bone
B. Percentage of body Ca not present in bone
Ref. 1 - p. 887

1476. A. Normal serum Ca concentration
B. Normal serum phosphorus concentration
Ref. 1 - pp. 888-890

1477. A. Percentage of inorganic material in bone
B. Percentage of organic material in bone
Ref. 1 - p. 890

1478. A. Daily requirement of human adult for Fe
B. Daily requirement of human adult for Ca
Ref. 1 - p. 1005

1479. A. Phosphorus excretion via feces
B. Phosphorus excretion via urine
Ref. 1 - p. 890

ANSWER THE FOLLOWING QUESTIONS BY USING THE KEY OUTLINED BELOW:
A. If 1, 2 and 3 are correct
B. If 1 and 3 are correct
C. If 2 and 4 are correct
D. If only 4 is correct
E. If all four are correct

1480. KWASHIORKOR:
1. Is characterized by growth retardation, anemia, hypoproteinemia, edema, and fatty infiltration of the liver
2. The symptoms respond therapeutically to a high-protein diet containing considerable meat and milk products
3. Some of the symptoms reflect a methionine deficiency which can be duplicated by feeding ethionine, a methionine antagonist
4. Is a disease observed primarily in highly-developed industrial areas
Ref. 1 - pp. 1009-1010

1481. SEVERE PROTEIN DEFICIENCY CAUSES:
1. Negative nitrogen balance
2. Decrease of RNA and DNA in liver
3. Complete loss of xanthine oxidase in liver after 2 weeks of protein-free diet
4. Sharply increased activity of endocrine system
Ref. 1 - pp. 1009-1010

SECTION I - DIET, ABSORPTION AND DIGESTION

1482. IRON:
1. Contained in hemoglobin is excreted when the erythrocytes disintegrate
2. Is found in the plasma mainly bound to a globulin called transferrin
3. Is in the ferrous form in ferritin
4. Is deposited in tissues (siderosis) when iron intake is excessive, in pernicious anemia and in hemochromatosis

Ref. 1 - pp. 746-749

1483. FERRITIN, THE STORAGE FORM OF IRON, IS FOUND IN:
1. Liver
2. Intestinal mucosa
3. Spleen
4. Bone

Ref. 1 - pp. 748-749

1484.
1. Magnesium is not an essential element in the diet
2. Normal serum magnesium concentration is 15 to 30 mg. per 100 ml
3. 90% of magnesium in the serum is protein bound
4. Most of the magnesium in the body is in the skeleton

Ref. 1 - pp. 1014-1015

1485.
1. Ingestion of magnesium increases excretion of calcium
2. Magnesium deprivation does not produce detectable symptoms
3. Magnesium is essential for many enzymatic reactions of carbohydrate metabolism
4. Magnesium has no important function in plants

Ref. 1 - pp. 239;389;1014-1015

1486.
1. Animals and man can be maintained on diets devoid of carbohydrate
2. Lactose supplies the major portion of the caloric value of milk
3. Of the pentoses mainly ribose and deoxyribose are utilized by animals
4. Carbohydrate supplies about 10 to 30% of the calories of most human diets

Ref. 1 - pp. 1013-1014

1487.
1. The protein concentration of plants is higher than that of animals
2. Plant proteins are relatively lower in lysine, methionine and tryptophan than animal protein
3. Animals store amino acids as a reserve food
4. Animals synthesize protein efficiently only when the component amino acids are present simultaneously

Ref. 1 - pp. 1006-1009

1488. THE BIOLOGICAL VALUE OF A PROTEIN IS DETERMINED IN PART BY ITS:
1. Acidity or basicity
2. Digestibility
3. Molecular weight
4. Amino acid composition

Ref. 1 - pp. 1006-1008

1489. SELENIUM:
1. Can prevent acute, fatal hepatic necrosis in rats on a diet deficient in sulfur amino acids and vitamin E
2. In plants, selenium can replace S in sulfur-containing amino acids
3. Selenium in large doses is toxic
4. Selenium poisoning has never been a public health problem in the U. S.

Ref. 1 - p. 1017

MATCH THE FORMULAS WITH THE CORRECT NAME:

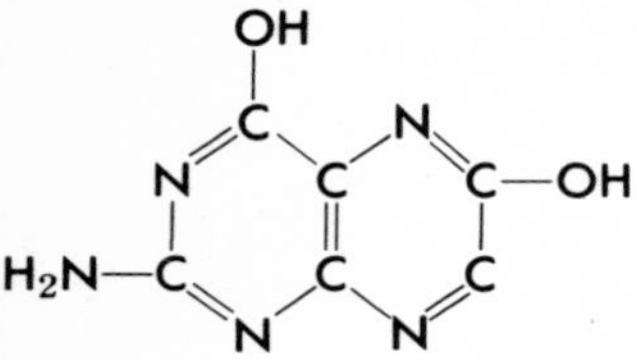

A.

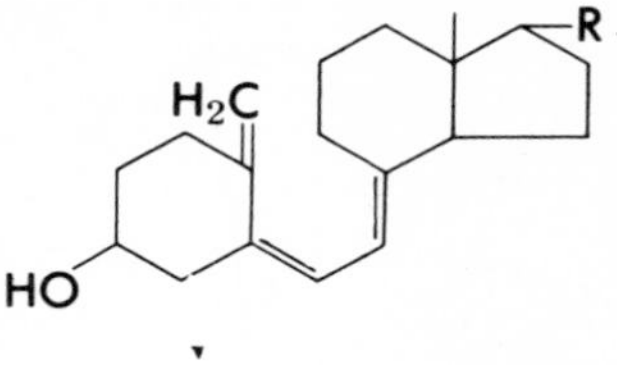

$R = C_8H_{17}$

B.

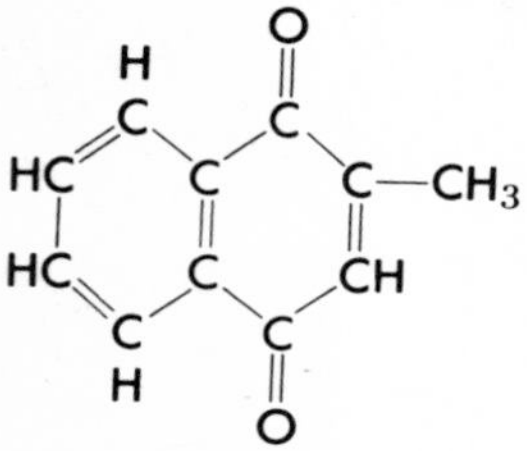

C.

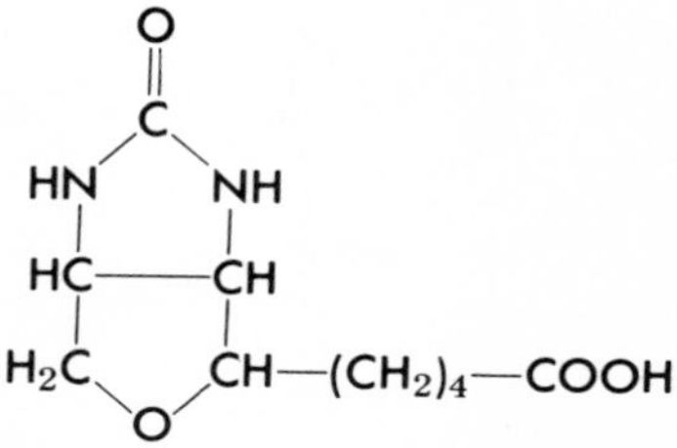

D.

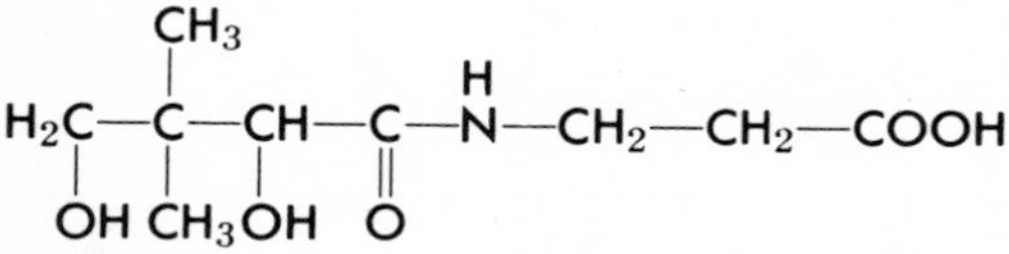

E.

1490. ___ Xanthopterin

1491. ___ Pantothenic acid

1492. ___ Menadione

1493. ___ Vitamin D

1494. ___ Oxybiotin

Ref. 1 - pp. 80; 1027; 1031; 1033; 1035

MATCH THE FORMULAS WITH THE CORRECT NAME:

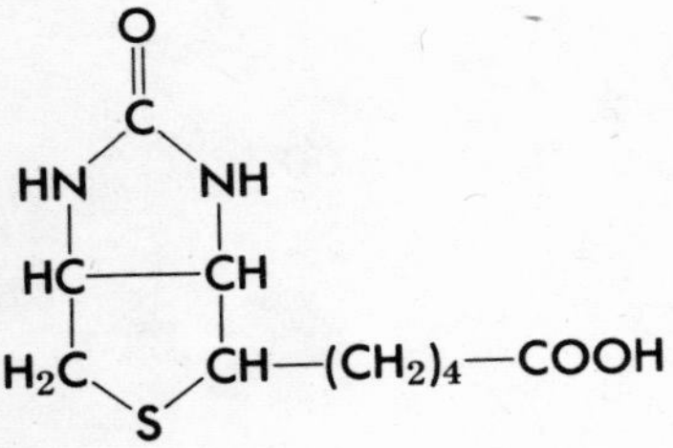

A.

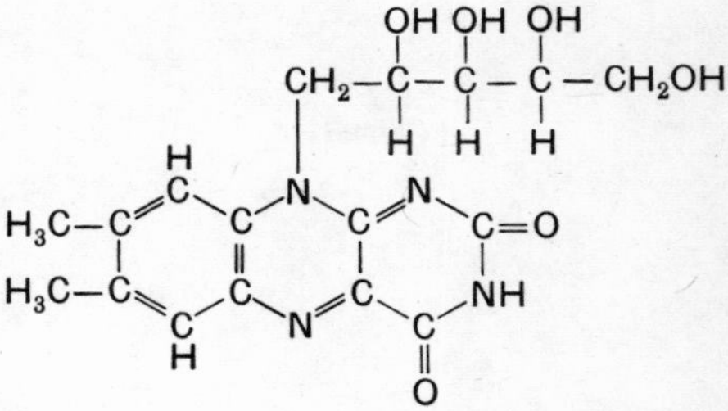

B.

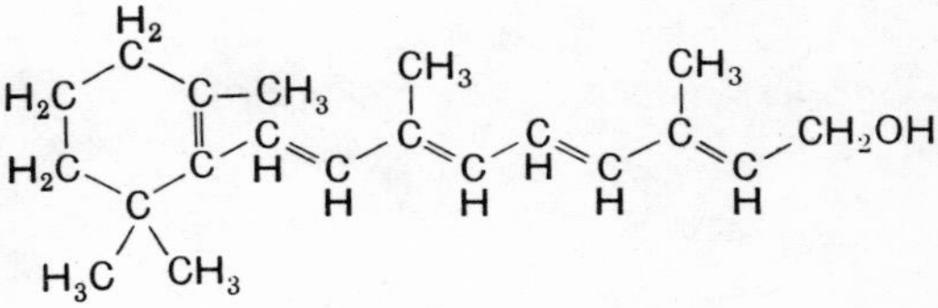

C.

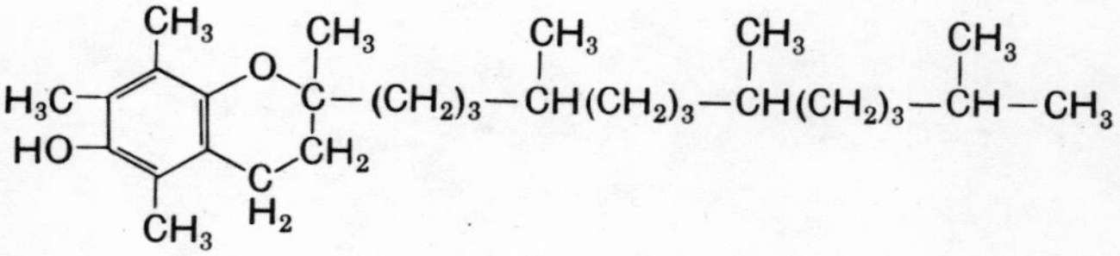

D.

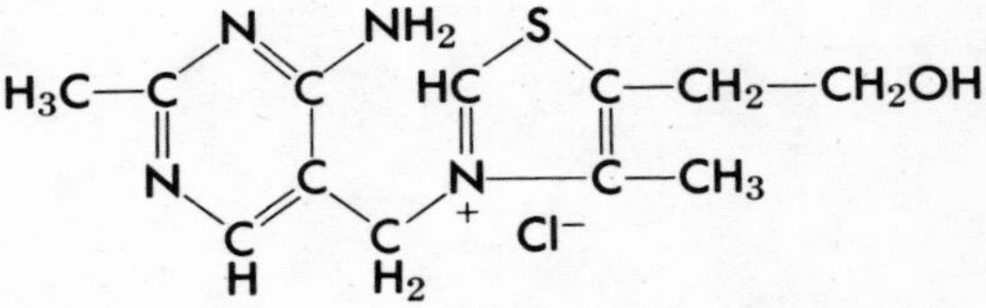

E.

1495. ___ Riboflavin

1496. ___ Thiamine

1497. ___ Biotin

1498. ___ α -Tocopherol

1499. ___ Vitamin A_1

Ref. 1 - pp. 992; 1020; 1023; 1033; 1054

MATCH THE FORMULAS WITH THE CORRECT STRUCTURE:

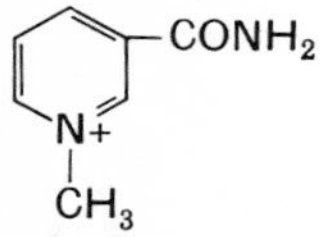

$CONH_2$, N^+, CH_3

A.

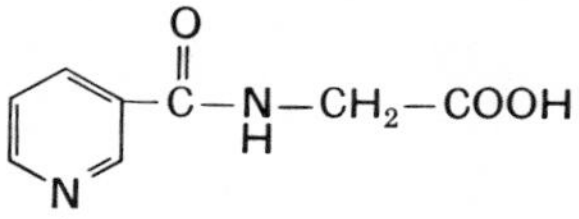

O, C, N, H, CH_2, COOH, N

B.

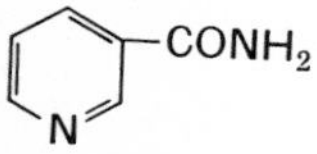

$CONH_2$, N

C.

COOH, N

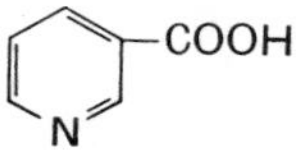

D.

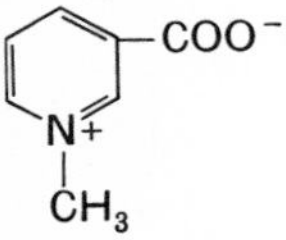

COO^-, N^+, CH_3

E.

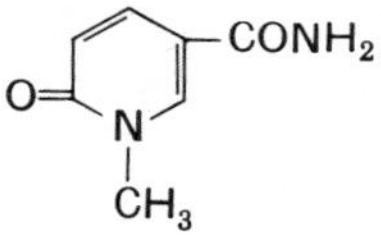

O, $CONH_2$, N, CH_3

F.

1500. ___ N-methylnicotinamide

1501. ___ Nicotinamide

1502. ___ Trigonelline

1503. ___ Nicotinuric acid

1504. ___ Niacin

1505. ___ N-Methyl-6-pyridone-3-carboxamide

Ref. 1 - p. 1027

ANSWER THE FOLLOWING QUESTIONS BY USING THE KEY OUTLINED BELOW:

A. If 1, 2 and 3 are correct
B. If 1 and 3 are correct
C. If 2 and 4 are correct
D. If only 4 is correct
E. If all four are correct

1506. VITAMIN A:
1. Has xerophthalmic activity
2. Is formed in the liver from various provitamins
3. Can be formed from parenterally injected lycopene
4. Is an isoprenoid alcohol

Ref. 1 - pp. 1049-1050

1507. VITAMIN C:
1. Is required for proline hydroxylation
2. Is a strong oxidizing agent
3. Is not required by albino rats
4. Is the coenzyme of p-hydroxyphenylpyruvic acid oxidase

Ref. 1 - pp. 545; 1044-1045

1508. RIBOFLAVIN IS A CONSTITUENT OF:
1. Cocarboxylase
2. Codecarboxylase
3. Diphosphopyridine nucleotide
4. FAD

Ref. 1 - pp. 641; 1023

1509. CHRONICALLY NIACIN DEFICIENT DOGS:
1. Exhibit profound anemia and hypoproteinemia
2. Exhibit a marked hematopoietic response if niacin is given but food consumption is restricted to what the dog had been eating previously
3. Exhibit an increase in serum proteins if niacin is given and there is a small increase in the food allowance
4. Are not able to increase body weight even if niacin and food are given ad libitum

Ref. 1 - pp. 1010; 1027-1028

1510. VITAMIN E:
1. Deficiency results in muscular dystrophy in rabbits
2. Deficient muscles contain abnormally large amounts of creatine
3. Has antioxidant activity for vitamin A
4. Is a co-factor of serum aldolase, explaining the fact that the activity of this enzyme is reduced in muscular dystrophy

Ref. 1 - p. 1055

1511. VITAMIN B_{12}:
1. As little as 3 micrograms give a positive hematologic response in man
2. Inhibits the dismutation of vicinal diols
3. Participates in the interconversion of succinyl CoA and methylmalonyl CoA
4. Pernicious anemia is due to inadequate ingestion of Vitamin B_{12}

Ref. 1 - pp. 1038-1041

1512. CONCERNING VITAMIN A:
1. Dietary status with respect to vitamin A can be evaluated by measurement of the visual threshold
2. When excessive vitamin A is ingested it accumulates in the liver where it is converted to vitamin A acid
3. In clinical vitamin A deficiency night blindness is one of the symptoms. This means that the visual threshold may be 100 times higher than normal
4. Vitamin A acid can cure xerophthalmia provided it is administered in time Ref. 1 - pp. 1049-1050

1513. THE STRUCTURE OF VITAMIN B_{12} CONTAINS:
1. Divalent cobalt bound to carbon
2. Ribose-3-phosphate
3. 5'-deoxyadenosine
4. 5, 6-dimethylbenzimidazole Ref. 1 - p. 1038

1514. CONCERNING VITAMIN B_6:
1. The pyridoxine requirement of animals varies directly with the protein content of the diet
2. Pyridoxal phosphate is probably an essential factor in the esterification of cholesterol with long-chain fatty acids
3. The conversion of serine to pyruvate by serine dehydrase requires pyridoxal phosphate
4. The symptoms of pyridoxine-deficiency can be reversed by administration of isoniazid Ref. 1 - pp. 495;597;1029-1030

1515. CONCERNING PERNICIOUS ANEMIA:
1. It is the consequence of inadequate intake of vitamin B_{12}
2. Vitamin B_{12} is identical with "intrinsic factor" of gastric juice
3. The hemopoietic failure in this disease is due to an overproduction of a specific mucoprotein
4. The gastric juice of patients with pernicious anemia contains no free HCl or pepsin Ref. 1 - pp. 752-753

EACH OF THE QUESTIONS OR INCOMPLETE STATEMENTS BELOW IS FOLLOWED BY 5 SUGGESTED ANSWERS OR COMPLETIONS. SELECT THE ONE WHICH IS BEST IN EACH CASE:

1516. THE BIOLOGICAL ACTIVITY OF THE TOCOPHEROLS HAS BEEN ATTRIBUTED IN PART TO THEIR ACTION AS:
A. Antioxidants
B. Coenzymes for transamination
C. Coenzymes in glycolysis
D. Precursors in the biosynthesis of vitamin A
E. Constituents of connective tissue Ref. 1 - p. 1056

1517. VITAMIN B_6 IS A COENZYME IN ALL OF THE FOLLOWING REACTIONS, EXCEPT:
A. Desulfuration of homocysteine
B. Transaminations
C. Racemization of optically active amino acids
D. Glutamate mutase reaction
E. Dehydration of serine Ref. 3 - pp. 394-400, 426

CHAPTER V - NUTRITION
SECTION II - VITAMINS

1518. WHICH OF THE FOLLOWING VITAMINS CONTAINS COBALT?:
A. Vitamin A
B. Vitamin B_1
C. Vitamin C
D. Vitamin B_6
E. Vitamin B_{12}
Ref. 1 - p. 1038

1519. VITAMIN K PLAYS AN ESSENTIAL ROLE IN:
A. Preventing thrombosis
B. The biosynthesis of prothrombin and proconvertin
C. Maintaining retinal integrity
D. Preventing bile stasis
E. Oxidative phosphorylation
Ref. 1 - pp. 731; 1058

1520. LIVER AND HEART MITOCHONDRIA CONTAIN A GROUP OF QUINONES WHICH ARE STRUCTURALLY RELATED TO THE K VITAMINS; ONE NAME FOR THESE COMPOUNDS IS:
A. Isoprene
B. Coenzyme Q
C. Menadione
D. Alpha-tocopherylquinone
E. Alpha-tocopherylhydroquinone
Ref. 1 - pp. 323; 340

1521. VITAMIN B_{12} IS A:
A. Porphyrin-like compound
B. Fat soluble vitamin
C. A copper-containing B-vitamin
D. Vitamin synthesized by all mammals except man
E. Co-enzyme in transamination
Ref. 1 - p. 1038

1522. AMINOPTERIN MAY PRODUCE TEMPORARY REMISSIONS OF ACUTE LEUKEMIA IN CHILDREN BY VIRTUE OF ITS:
A. General cytotoxic action
B. Action as an inhibitor of cell respiration
C. Action as an antimetabolite of folic acid
D. Inhibition of protein synthesis
E. Folic acid activity
Ref. 1 - p. 1037

1523. THE "XANTHURENIC ACID INDEX" CAN BE USED AS A MEASURE OF PYRIDOXINE DEFICIENCY. IT INVOLVES THE METABOLISM OF THE AMINO ACID:
A. Histidine
B. Phenylalanine
C. Tryptophan
D. Serine
E. Methionine
Ref. 1 - pp. 615; 1030

1524. INJECTION OF VITAMIN B_{12} IN PATIENTS WITH PERNICIOUS ANEMIA OVERCOMES:
A. Lack of intrinsic factor
B. Lack of extrinsic factor
C. Hydrochloric acid production
D. Folic acid deficiency
E. Iron deficiency
Ref. 1 - pp. 1039-1040

1525. WHICH OF THE FOLLOWING MAJOR METABOLIC PROCESSES DOES NOT LEAD TO THE GENERATION OF ATP?:
A. Pyruvate → acetyl CoA
B. Glucose → pyruvate
C. Phosphocreatine → creatine
D. Oxaloacetate → glucose
E. Fatty acid → acetyl CoA
Ref. 5 - p. 537

1526. THE FEEDING OF AVIDIN, A BASIC PROTEIN PRESENT IN RAW EGG WHITE, MAY RESULT IN A DEFICIENCY OF:
A. Riboflavin
B. Vitamin A
C. Choline
D. Biotin
E. Nicotinamide
Ref. 1 - pp. 1033-1034

1527. ALL OF THE FOLLOWING VITAMINS ARE FAT SOLUBLE, EXCEPT:
A. Vitamin A
B. Vitamin C
C. Vitamin D
D. Vitamin K
E. Vitamin E
Ref. 1 - p. 1044

1528. THE COENZYME OF THE "OLD YELLOW ENZYME" CONTAINS:
A. Riboflavin
B. Thiamine
C. Biotin
D. Folic acid
E. Niacin
Ref. 1 - p. 1023

1529. BETA-CAROTENE IS MAINLY CONVERTED TO VITAMIN A IN THE:
A. Liver
B. Intestine
C. Spleen
D. Adrenal cortex
E. Kidney
Ref. 1 - p. 1049

1530. THE PYRIDINE NUCLEOTIDES, DPN AND TPN, ARE ESSENTIAL CO-FACTORS IN THE METABOLISM OF MANY SUBSTANCES. AN IMPORTANT PART OF THE STRUCTURE OF THESE MOLECULES IS:
A. Vitamin A
B. Vitamin C
C. Nicotinamide
D. Vitamin E
E. Pantothenic acid
Ref. 1 - p. 1026

1531. A VITAMIN WHICH IS A STRONG REDUCING AGENT, A PROPERTY WHICH MAY EXPLAIN ITS FUNCTION, IS:
A. Niacinamide
B. Thiamine
C. Vitamin B_2
D. Vitamin C
E. Folic acid
Ref. 1 - p. 1044

1532. BIOCHEMICAL ROLE OF THE CAROTENOIDS IS THEIR CONVERSION TO:
A. Folic acid
B. Folinic acid
C. Ascorbic acid
D. Vitamin A
E. Arachidonic acid
Ref. 1 - pp. 1048-1049

1533. THE GROWTH OF BACTERIA REQUIRING p-AMINOBENZOIC ACID IS INHIBITED BY:
A. Folic acid
B. Tetrahydrofolic acid
C. Citrovorum factor
D. Penicillin
E. Sulfonamides
Ref. 1 - pp. 240-241

1534. VITAMIN D IS REQUIRED FOR THE PREVENTION OF:
A. Beriberi, dry form
B. Scurvy
C. Rickets
D. Beriberi, wet form
E. Pellagra
Ref. 1 - p. 1052

1535. WHOLE WHEAT IS AN EXCELLENT SOURCE OF:
A. Vitamin D
B. Ascorbic acid
C. Thiamine
D. Vitamin A
E. All of these
Ref. 1 - p. 1022

1536. THE PELLAGRA-PREVENTIVE FACTOR IN THE VITAMIN B COMPLEX IS:
A. Pantothenic acid
B. Vitamin B_{12}
C. Thiamine
D. Niacin
E. Pyridoxine
Ref. 1 - p. 1025

1537. A BETAINE WHICH IS FORMED FROM NICOTINIC ACID IN PLANTS IS:
A. Riboflavin phosphate
B. Trigonelline
C. Thiamine pyrophosphate
D. Carnitine
E. Betaine aldehyde
Ref. 1 - pp. 1026-1027

1538. WHICH OF THE FOLLOWING COMPOUNDS IS NOT A METABOLITE OF NIACIN?:
A. N'-methyl-6-pyridone-3-carboxamide
B. Nicotinuric acid
C. Trigonelline
D. p-Aminobenzoic acid
E. N'-methylnicotinamide
Ref. 1 - p. 1027

1539. TRYPTOPHAN IS A PRECURSOR OF NICOTINIC ACID IN MAN. THERE IS NEVERTHELESS A REQUIREMENT FOR NIACIN IN MAN BECAUSE:
A. The conversion of nicotinic acid to niacin is too slow
B. The protein in corn has a low tryptophan content
C. The rate of niacin synthesis from proteins is too slow
D. Niacin derived from tryptophan is not available for DPN biosynthesis
E. The tryptophan was not given in the form of protein
Ref. 1 - pp. 1027-1028

EACH GROUP OF QUESTIONS BELOW CONSISTS OF SEVERAL LETTERED HEADINGS FOLLOWED BY A LIST OF NUMBERED WORDS OR PHRASES. FOR EACH NUMBERED WORD OR PHRASE SELECT THE ONE LETTERED HEADING THAT IS MOST CLOSELY RELATED TO IT:

MATCH THE INORGANIC BODY CONSTITUENTS WITH THE CLINICAL MANIFESTATION OF THEIR DEFICIENCY:

A. Calcium
B. Potassium
C. Fluoride
D. Copper
E. Magnesium

1540. ___ Hypochromic anemia
1541. ___ Rickets
1542. ___ Muscular tremors, hyperirritability
1543. ___ Dental caries
1544. ___ Acidosis

Ref. 2 - pp. 436-439

MATCH THE VITAMINS LISTED BELOW WITH THE CLINICAL MANIFESTATION OF THEIR DEFICIENCY:

A. Cobalamin
B. Ascorbic acid
C. Niacin
D. Folic acid
E. Thiamine

1545. ___ Scurvy
1546. ___ Pernicious anemia, sprue
1547. ___ Macrocytic anemia, sprue
1548. ___ Beriberi
1549. ___ Pellagra

Ref. 2 - pp. 830-831

MATCH THE VITAMINS LISTED BELOW WITH THEIR PRINCIPAL METABOLIC FUNCTIONS:

A. Retinol
B. Cholecalciferol
C. D- α-tocopherol
D. Phylloquinone
E. Pyridoxal phosphate

1550. ___ Coenzyme of transaminases
1551. ___ Intestinal absorption of phosphate
1552. ___ Maintenance of epithelial tissue
1553. ___ Antioxidant
1554. ___ Essential for prothrombin formation

Ref. 2 - pp. 813, 792

A. Vitamin D
B. Alpha-tocopherol
C. Vitamin K
D. Vitamin A
E. Inositol

1555. ___ Earliest sign of deficiency is night blindness
1556. ___ Mediates calcium absorption from intestine
1557. ___ Deficiency in rats causes decrease in muscle creatine, creatinuria, and slight fall in creatinine excretion
1558. ___ Plays a role in blood clotting
1559. ___ A vitamin which can be synthesized from glucose in the rat

Ref. 1 - pp. 1042, 1050, 1053, 1055, 1057-1058

A. Possesses vitamin activity
B. Possesses antivitamin activity due to structural similarity
C. Inactivates the vitamin but is not a structural analogue
D. Has no known vitamin activity

1560. ___ Pyridine-3-sulfonic acid
1561. ___ Oxybiotin
1562. ___ Avidin
1563. ___ Nicotinic acid

Ref. 1 - pp. 241, 1025, 1033-1034

A. Possesses vitamin activity
B. Possesses antivitamin activity due to structural similarity
C. Inactivates the vitamin but is not a structural analogue
D. Has no known vitamin activity

1564. ___ Isonicotinylhydrazide
1565. ___ Aminopterin
1566. ___ 2-keto-L-gulonolactone
1567. ___ Tachysterol Ref. 1 - pp. 1030, 1035, 1045, 1052

DEFICIENCY OF:
A. Ascorbic acid
B. Thiamine
C. Niacin
D. Pyridoxine
E. Calciferol

PRODUCES:
1568. ___ Pellagra
1569. ___ Rickets
1570. ___ Acrodynia (in rats)
1571. ___ Scurvy
1572. ___ Beriberi Ref. 1 - pp. 1020, 1026, 1028, 1045, 1052

ANSWER THE FOLLOWING QUESTIONS BY USING THE KEY OUTLINED BELOW:
A. If A is greater than B
B. If B is greater than A
C. If A and B are equal or nearly equal

1573. A. Recommended daily allowance of vitamin C in mg
B. Recommended daily allowance of vitamin A in mg
Ref. 1 - pp. 1005, 1048

1574. A. Vitamin A activity of 1 molecule of alpha-carotene
B. Vitamin A activity of 1 molecule of beta-carotene
Ref. 1 - p. 1048

1575. A. Fecal excretion of orally administered vitamin B_{12} in normal man
B. Fecal excretion of orally administered vitamin B_{12} in pernicious anemia patient Ref. 1 - p. 1039

1576. A. Conversion of L-gulono-gamma-lactone to ascorbic acid in guinea pigs
B. Conversion of L-gulono-gamma-lactone to ascorbic acid in man
Ref. 1 - p. 1044

1577. A. Toxic effect of overdose of vitamin A in man
B. Toxic effect of overdose of vitamin C in man
Ref. 1 - pp. 1046, 1051

1578. A. Weight of one I. U. of vitamin D
B. Weight of one I. U. of vitamin A Ref. 1 - pp. 1048, 1052

MATCH THE STRUCTURES WITH THE CORRECT NAME:

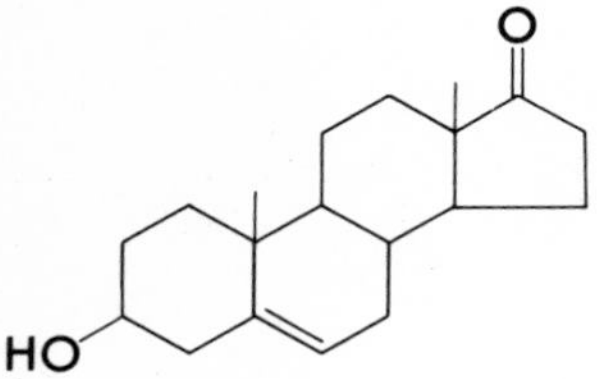

A.

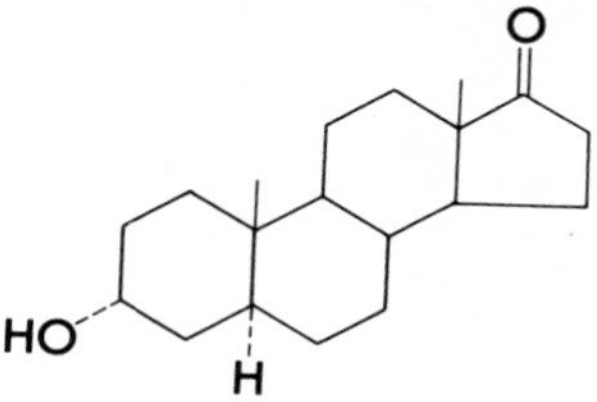

B.

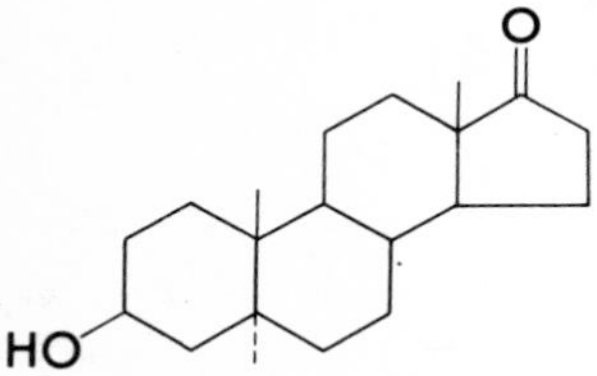

C.

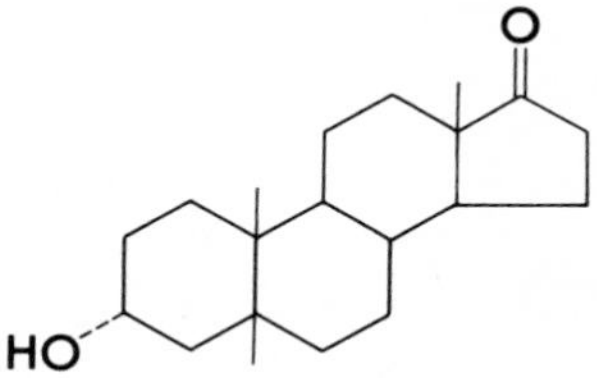

D.

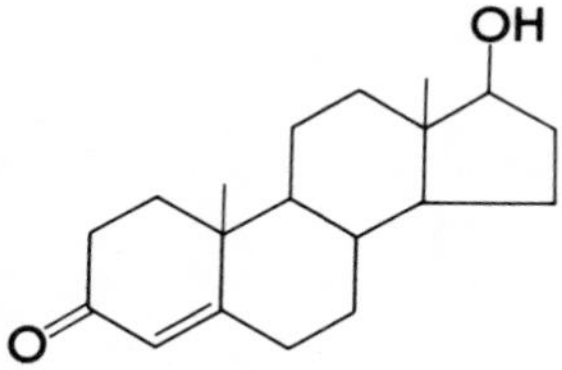

E.

1579. ___ Dehydroepiandrosterone

1580. ___ Testosterone

1581. ___ Androsterone

1582. ___ Etiocholan – 3 α -ol-17-one

1583. ___ Epiandrosterone

Ref. 1 - pp. 937; 940

MATCH THE STRUCTURES WITH THE CORRECT NAME:

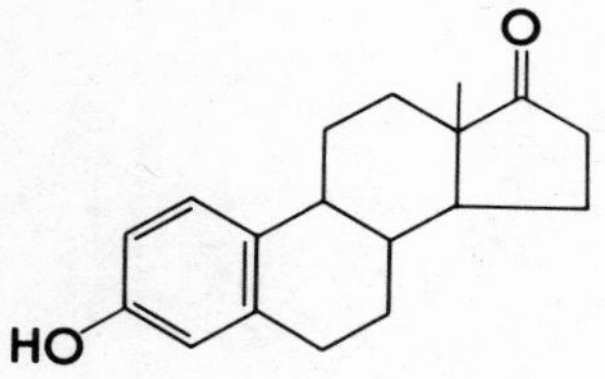

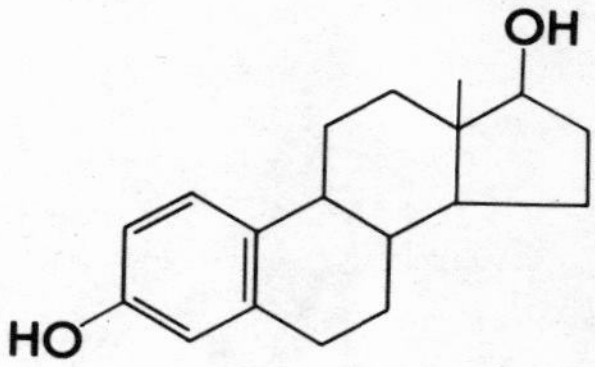

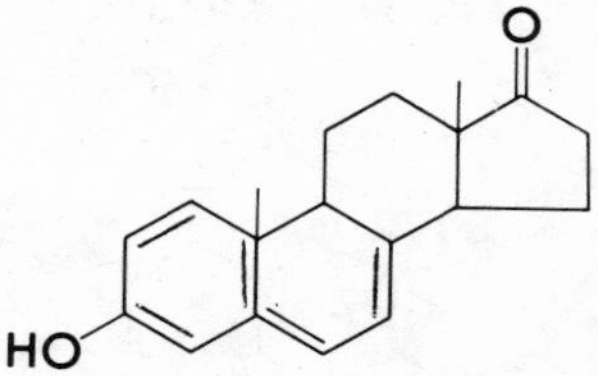

1584. ___ Equilin

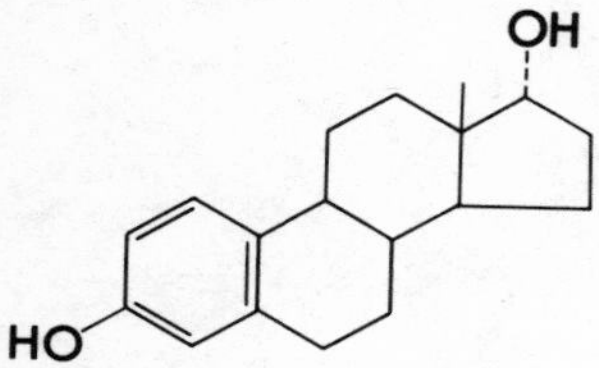

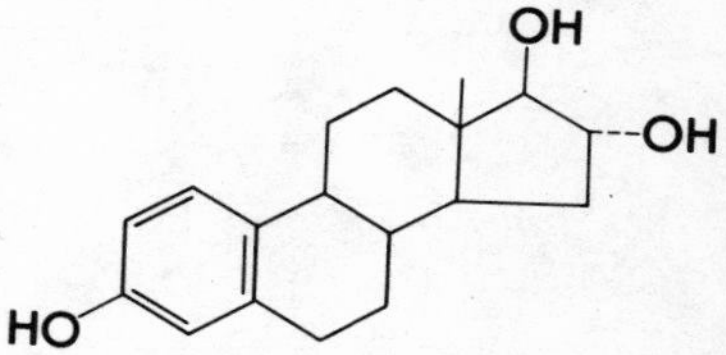

1585. ___ α-Estradiol

1586. ___ Estriol

1587. ___ Estrone

1588. ___ β -Estradiol

Ref. 1 - pp. 944-945

MATCH THE STRUCTURES WITH THE CORRECT NAME:

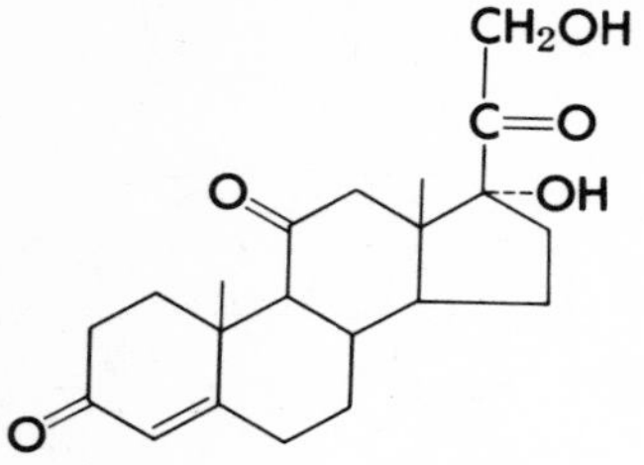

A.

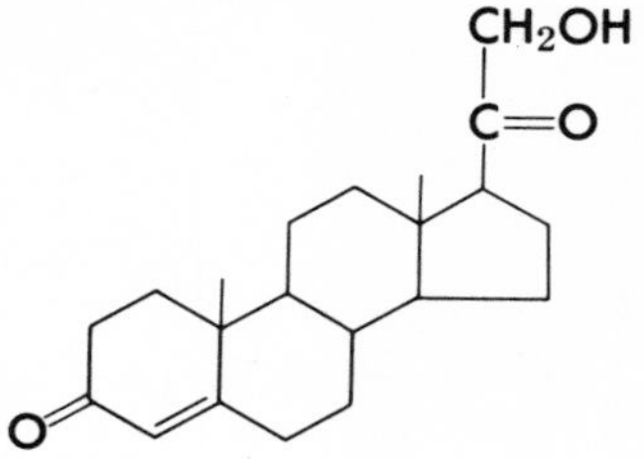

B.

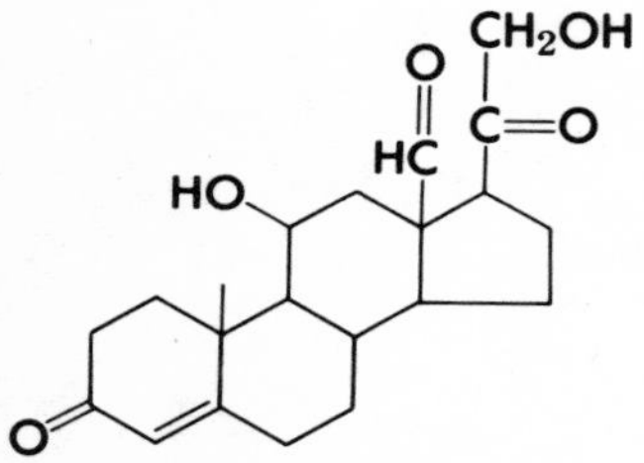

C.

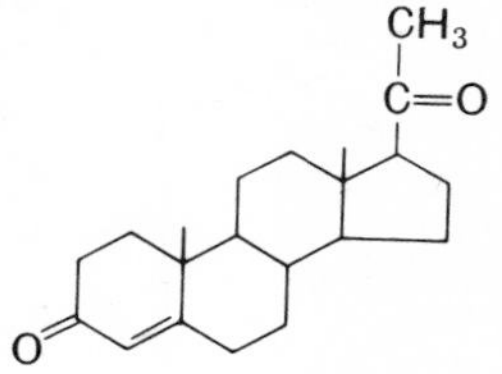

D.

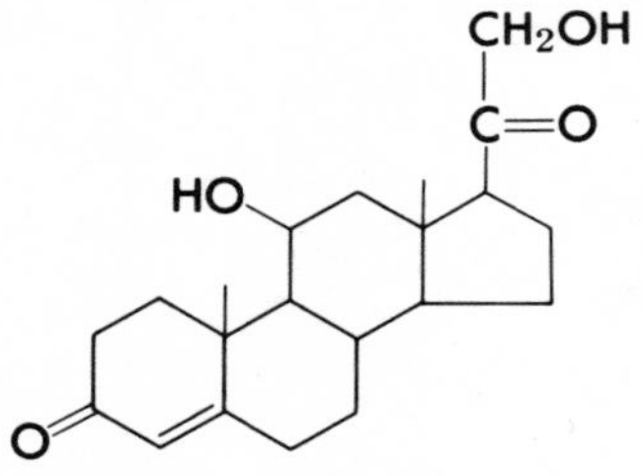

E.

1589. ___ Cortisone

1590. ___ Deoxycorticosterone

1591. ___ Corticosterone

1592. ___ Progesterone

1593. ___ Aldosterone

Ref. 1 - pp. 958-960

EACH OF THE QUESTIONS OR INCOMPLETE STATEMENTS BELOW IS FOLLOWED BY 5 SUGGESTED ANSWERS AND COMPLETIONS. SELECT THE ONE WHICH IS BEST IN EACH CASE:

1594. THE PARATHYROID GLAND REGULATES THE METABOLISM OF:
A. Calcium
B. Phosphate
C. Both A and B
D. Calcium and magnesium
E. Phosphate and sodium
Ref. 1 - pp. 932-935

1595. A HYPERGLYCEMIC FACTOR PRODUCED BY THE PANCREAS IS:
A. Insulin
B. Lipase
C. Glucagon
D. FSH
E. Cathepsin
Ref. 1 - pp. 977-978

1596. GIGANTISM AND ACROMEGALY ARE DISEASES DUE TO ALTERATIONS IN SECRETION OF WHICH HORMONE?:
A. FSH
B. TSH
C. Somatotrophin
D. Parathormone
E. Melatonin
Ref. 1 - pp. 995-997

1597. A SUBSTANCE PRESENT IN THE SMALL INTESTINE WHICH STIMULATES CONTRACTION OF THE GALL BLADDER IS:
A. Enterocrinin
B. Secretin
C. Cholecystokinin
D. Enterokinase
E. Intrinsic factor
Ref. 1 - pp. 818-819

1598. THE HORMONE WHICH IS ESSENTIAL FOR THE METABOLISM OF GLUCOSE IS CHARACTERIZED BY THE POSSESSION OF:
A. Two peptide chains joined by disulfide bonds
B. Two peptide chains joined by carbon to nitrogen bonds
C. 14 amino acids
D. One peptide chain consisting of 10 amino acid residues
E. D-amino acids
Ref. 1 - pp. 147-149

1599. THE HORMONE THYROXINE IS SYNTHESIZED IN THE ORGANISM FROM:
A. Tyrosine
B. Indole-5, 6-quinone
C. L-histidine
D. Tyramine
E. Thyroglobulin
Ref. 1 - pp. 922-923

1600. IN MAMMALS, NOREPINEPHRINE IS SYNTHESIZED FROM:
A. Pyruvate
B. Arginine
C. Catechol
D. Tyrosine
E. Tryptamine
Ref. 1 - p. 954

1601. PROGESTERONE IS A PRECURSOR OF THE FOLLOWING BIOLOGICALLY IMPORTANT STEROIDS:
A. Aldosterone, cortisone, corticosterone
B. Cholesterol, cortisone, cholic acid
C. Aldosterone, deoxycholic acid, pregnenolone
D. Hydrocortisone, pregnenolone, estriol
E. Corticosterone, testosterone, cholesterol sulfate
Ref. 2 - p. 469

1602. IN DIABETES THE ACTIVITY OF CERTAIN HEPATIC ENZYMES IS INCREASED. WHICH OF THE FOLLOWING HEPETIC ENZYME ACTIVITIES IS NOT INCREASED IN THE DIABETIC STATE ?:
A. CoA-carnitine acyl transferase
B. Glucose-6-phosphatase
C. Pyruvate carboxylase
D. Serine dehydratase
E. Citrate cleavage enzyme Ref. 5 - p. 561

1603. STEROIDS WITH 8 TO 10 CARBON ATOMS IN THE SIDE CHAIN AT POSITION 17 AND AN ALCOHOLIC HYDROXYL GROUP AT POSITION 3 ARE CLASSIFIED AS:
A. Sterols
B. Bile acids
C. Adrenal cortical steroids
D. Androgens
E. Estrogens Ref. 1 - p. 79

1604. THE MECHANISM OF ACTION OF INSULIN AT THE CELLULAR LEVEL IS NOT KNOWN WITH CERTAINTY. TWO CURRENTLY HELD VIEWS ON INSULIN ACTION ARE:
A. Insulin promotes the conversion of fat to carbohydrate and facilitates the passage of glucose into the cell
B. Insulin facilitates the passage of glucose into the cell and inhibits fatty acid oxidation
C. Insulin accelerates the metabolic utilization of fructose-6-phosphate and stimulates hexokinase activity
D. Insulin stimulates hexokinase activity and promotes the production of epinephrine
E. Insulin stimulates hexokinase activity and facilitates the passage of glucose into the cell Ref. 1 - pp. 974-975

1605. IN INSULIN DEFICIENCY:
A. Protein synthesis is depressed
B. Protein degradation is increased
C. Nitrogen excretion is increased
D. Fatty acid synthesis is depressed
E. All of the above apply Ref. 1 - pp. 448; 975

1606. IN DIABETIC SUBJECTS PREVIOUSLY TREATED WITH INSULIN, INTRAVENOUSLY ADMINISTERED INSULIN IS DESTROYED LESS RAPIDLY, AND GIVES A LESSER RESPONSE, THAN IN NORMAL SUBJECTS. THIS OBSERVATION:
A. Is untrue
B. Is due to the presence of a non-precipitating insulin antibody in the blood of the diabetic subjects
C. Is explained in terms of an increase of insulinase activity in the blood of diabetics
D. Is ascribed to an effect of somatotrophin since this hormone can raise the level of blood glucose
E. First suggested that insulin can separate into A and B chains in vivo
Ref. 1 - p. 977

1607. GLUCAGON AND EPINEPHRINE ARE SIMILAR IN THAT:
A. Both promote glycogenolysis in liver and muscle
B. Both promote the reactivation of liver phosphorylase phosphokinase
C. Both reduce blood glucose levels
D. Both are formed by the alpha-cells of the pancreas
E. Both are hypertensive agents Ref. 1 - pp. 440; 977

1608. THE PROSTAGLANDINS:
A. Cause hypertension
B. Occur only in prostatic tissue
C. Are alicyclic fatty acid derivatives
D. Are synthesized biologically from oleic acid
E. Increase irritability of nerve tissue
Ref. 1 - pp. 943-944

1609. IN HYPOPARATHYROIDISM IN MAN THERE OCCURS:
A. Elevation of serum calcium
B. Urinary calculi
C. Increased urinary excretion of calcium
D. Decrease in ionic serum calcium
E. None of these
Ref. 1 - pp. 934-935

1610. CHEMICALLY, THE ESTROGENS ARE CHARACTERIZED BY:
A. The 21 carbon steroid skeleton
B. The lack of a methyl group at C-10
C. The aromatic character of the D ring
D. The lack of an oxygen function at C-17
E. All of these
Ref. 1 - p. 944

1611. WHAT SUBSTANCE SERVES AS A PRECURSOR IN THE BIOSYNTHESIS OF TESTOSTERONE?:
A. Adrenosterone
B. Pregnenolone
C. Estrone
D. Methyltestosterone
E. Aldosterone
Ref. 1 - p. 939

1612. THE ANDROGENS ARE EXCRETED IN THE URINE LARGELY:
A. As estradiol
B. Unchanged
C. As glucuronic acid and sulfate conjugates
D. As phenolic sulfate esters
E. As 16 α-hydroxy compounds
Ref. 1 - p. 941

1613. FOLLICLE STIMULATING HORMONE AND INTERSTITIAL CELL-STIMULATING HORMONE ARE PRODUCED BY:
A. Alpha cells of pancreas
B. Adenohypophysis
C. Follicle
D. Cells of Leydig
E. Placenta
Ref. 1 - p. 919

1614. AN EXAMPLE OF A STEROID CONTAINING AN AROMATIC RING:
A. Progesterone
B. Testosterone
C. Estrone
D. Cortisone
E. Estrane
Ref. 1 - p. 914

1615. CHEMICALLY, STEROIDS ARE DERIVATIVES OF:
A. Fatty acids
B. Cholesterol
C. Ergosterol
D. Perhydrocyclopentanophenanthrene
E. None of these
Ref. 1 - p. 77

1616. IN THE BIOSYNTHESIS OF ESTROGENS:
A. Androgens serve as precursors
B. The angular methyl group at C-10 is lost
C. An aromatization reaction occurs which requires O_2 and TPNH
D. Phenolic substances are the end-products
E. All of the above apply
Ref. 1 - p. 945

1617. THIOURACIL INHIBITS THYROID FUNCTION. THIS EFFECT IS THOUGHT TO BE DUE TO:
A. Inhibition of synthesis of thyroid hormone
B. Hyperplasia of the thyroid gland
C. Reduction of TSH production by the pituitary
D. Formation of thiocyanate from thiouracil
E. Formation of L-thyroxine which acts as an antimetabolite
Ref. 1 - pp. 930-931

1618. MEASUREMENT OF URINARY NEUTRAL "17-KETOSTEROIDS":
A. Is done by means of the Porter-Silber method
B. Gives abnormally elevated values in children
C. Is used as a means of evaluating adrenal and testicular function
D. Gives abnormally low values in adrenocortical hyperplasia
E. Gives abnormally high values in hypogonadism
Ref. 1 - p. 942

1619. ALDOSTERONE:
A. Is the most active "mineralo-corticoid" discovered so far
B. Is probably a metabolite of dehydroepiandrosterone
C. Acts by preventing tubular reabsorption of sodium
D. Is a precursor of 11-deoxycorticosterone
E. Has an aldehyde function at C-17
Ref. 1 - p. 962

1620. THE C_{21}- ADRENAL CORTICAL STEROIDS ARE SYNTHESIZED IN VIVO FROM:
A. Testosterone
B. 18-hydroxycorticosterone
C. Cortisone
D. Progesterone
E. Aldosterone
Ref. 1 - p. 960

EACH GROUP OF QUESTIONS BELOW CONSISTS OF 5 LETTERED HEADINGS FOLLOWED BY A LIST OF NUMBERED WORDS OR PHRASES. FOR EACH NUMBERED WORD OR PHRASE SELECT THE ONE LETTERED HEADING THAT IS MOST CLOSELY RELATED TO IT:

A. Arylsulfonylurea derivatives
B. Epinephrine
C. Thyroxine
D. Glucagon
E. Parathyroid hormone

1621.___ A polypeptide,elaborated by pancreatic alpha-cells,which increases blood sugar
1622.___ Increase(s) insulin release by pancreas
1623.___ Enhance(s) protein synthesis in microsomes
1624.___ Cause(s) release of calcium from bone
1625.___ Increase(s) glycogenolysis in liver and muscle
Ref. 1 - pp. 926; 933; 957; 973; 977

ITEMS A TO E ARE FRAGMENTS OF STEROIDS. RELATE THESE FRAGMENTS CORRECTLY TO THE NUMBERED ITEMS 1626-1630 BELOW:

A. HO

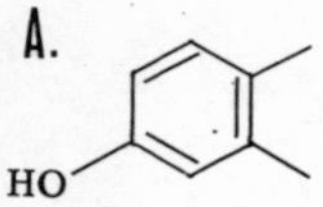

B. HO

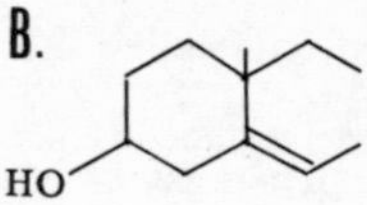

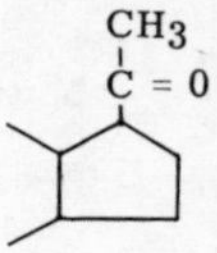

C. CH_2OH C = O OH

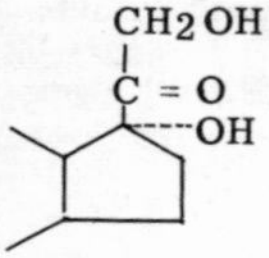

D. O

E. CH_3 C = 0

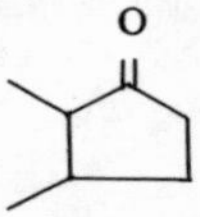

1626. ___ Progestational steroid
1627. ___ Estrogen
1628. ___ Many have an oxygen function at C-11
1629. ___ Determined by means of Zimmerman reaction (m-dinitrobenzene)
1630. ___ Intermediate in the formation of progesterone

Ref. 1 - pp. 939;942;944;950; 958;960

RELATE THE FOLLOWING HORMONES TO THE ENDOCRINE GLAND WHICH ELABORATES THEM:

A. Adenohypophysis
B. Neurohypophysis
C. Pars intermedia

1631. ___ MSH
1632. ___ Antidiuretic hormone
1633. ___ FSH
1634. ___ Prolactin
1635. ___ Oxytocin

Ref. 1 - pp. 919-920

RELATE THE FOLLOWING HORMONES AND THEIR FUNCTIONS:

A. Progesterone
B. ACTH
C. Somatotropin
D. Oxytocin
E. Relaxin

1636. ___ Release of lipid from adipose tissue
1637. ___ Muscle tone of symphysis pubis
1638. ___ Contraction of uterine smooth muscle
1639. ___ Development of alveolar system of mammary glands
1640. ___ Anabolic effect on P and N metabolism

Ref. 1 - pp. 918-920

A. FSH
B. Melanocyte stimulating hormone
C. ACTH
D. Insulin
E. Oxytocin

1641. ___ Causes lipid mobilization from adipose tissue
1642. ___ A glycoprotein
1643. ___ Function in man is unknown
1644. ___ Structure determined by duVigneaud
1645. ___ Promotes hepatic protein synthesis

Ref. 1 - pp. 975; 982; 985; 991; 993

A. Hydroxytyramine
B. 3-Methoxyepinephrine
C. S-Adenosylmethionine
D. Cyclic AMP
E. 3-Methoxy-4-hydroxymandelic acid

1646. ___ Precursor of norepinephrine
1647. ___ Required for conversion of norepinephrine to epinephrine
1648. ___ Metabolite of norepinephrine and epinephrine
1649. ___ Produced by action of catechol O-methyl transferase
1650. ___ Formation stimulated by epinephrine

Ref. 1 - pp. 954; 957

RELATE THE HORMONES LISTED BELOW TO THEIR PRINCIPAL SITES OF ACTION:

A. Calcitonin
B. Prostaglandin
C. Prolactin
D. ACTH
E. Pancreozymin

1651. ___ Mammary gland
1652. ___ Adrenal cortex
1653. ___ Gall bladder
1654. ___ Skeleton
1655. ___ Smooth muscle

Ref. 1 - pp. 917-920

ANSWER THE FOLLOWING GROUP OF QUESTIONS BY USING THE KEY OUTLINED BELOW:

A. If the item is associated with A only
B. If the item is associated with B only
C. If the item is associated with both A and B
D. If the item is associated with neither A nor B

A. Insulin
B. Epinephrine
C. Both
D. Neither

1656. ___ Promotes formation of fatty acids from carbohydrate
1657. ___ Secretion regulated in part by blood glucose level
1658. ___ Promotes conversion of inactive liver phosphorylase to active enzyme
1659. ___ Increases glycogenesis
1660. ___ Increases glycogenolysis

Ref. 1 - pp. 444; 957; 975

A. Hypothyroidism
B. Hyperthyroidism
C. Both
D. Neither

1661. ___ 1 to 2 micrograms of protein-bound iodine per 100 ml. blood
1662. ___ 4 to 8 micrograms of protein-bound iodine per 100 ml. blood
1663. ___ Elevated blood cholesterol
1664. ___ Negative nitrogen balance
1665. ___ Increased rate of iodine uptake by thyroid

Ref. 1 - pp. 926-928

A. Hormones of adenohypophysis
B. Hormones of neurohypophysis
C. Both
D. Neither

1666. ___ Nonapeptide amides
1667. ___ Polypeptides or proteins
1668. ___ Secretion influenced by osmotic pressure of blood
1669. ___ Trophic hormones
1670. ___ Also present in pineal gland

Ref. 1 - pp. 698, 981, 983, 985

A. Epinephrine
B. Norepinephrine
C. Both
D. Neither

1671. ___ Inhibits adenyl cyclase
1672. ___ Produces increase in blood non-esterified fatty acids
1673. ___ Causes increase in heart rate
1674. ___ Produces increase in total peripheral resistance
1675. ___ Produces increase in systolic blood pressure
1676. ___ Effect on adipose tissue similar to prostaglandins
1677. ___ Greatly reduced plasma level in pheochromocytoma

Ref. 1 - pp. 953-957

A. Oxytocin
B. Vasopressin
C. Both
D. Neither

1678. ___ Produced by adenohypophysis
1679. ___ Analogous to vasotocin of non-mammalian vertebrates
1680. ___ Controls polyuria and polydipsia of diabetes insipidus
1681. ___ Produces slight dilatation of coronary arteries
1682. ___ Stimulates intestinal contraction Ref. 1 - pp. 981-984

ANSWER THE FOLLOWING QUESTIONS BY USING THE KEY OUTLINED BELOW:
A. If A is greater than B
B. If B is greater than A
C. If A and B are equal or nearly equal

1683. A. $[K^+]$ of serum from adrenalectomized animals
B. $[K^+]$ of serum from normal animals
Ref. 1 - pp. 963; 965

1684. A. Extracellular $[Na^+]$ of brain from normal animals
B. Extracellular $[Na^+]$ of brain from adrenalectomized animals
Ref. 1 - p. 965

1685. A. Extracellular water in muscle from normal animals
B. Extracellular water in muscle from adrenalectomized animals
Ref. 1 - p. 965

1686. A. Urine content of 3-methoxy-4-hydroxymandelic acid in normal man
B. Urine content of 3-methoxy-4-hydroxymandelic acid in patient with pheochromocytoma Ref. 1 - p. 957

1687. TUBULAR REABSORPTION OF Na^+ + Cl^- BY KIDNEY:
A. In adrenal cortical insufficiency
B. In normal subjects Ref. 1 - p. 963

1688. A. Binding of aldosterone by transcortin
B. Binding of cortisol by albumin
Ref. 1 - p. 961

1689. A. Number of amino acids in oxytocin
B. Number of amino acids in vasopressin
Ref. 1 - p. 698

1690. A. Hyperglycemic effect of epinephrine
B. Hyperglycemic effect of norepinephrine
Ref. 1 - p. 955

1691. A. Serum cholesterol after thiouracil administration
B. Serum cholesterol after thyroxine administration
Ref. 1 - pp. 926-928

1692. A. Potency of orally administered testosterone
B. Potency of parenterally administered testosterone
Ref. 1 - p. 942

1693. A. Effect of cortisol on carbohydrate metabolism
B. Effect of aldosterone on carbohydrate metabolism
Ref. 1 - pp. 962-965

1694. A. Adrenal function in Cushing's disease
B. Adrenal function in normal subject Ref. 1 - p. 968

1695. A. Relative androgenic activity of testosterone
B. Relative androgenic activity of androsterone
Ref. 1 - p. 941

1696. A. Relative estrogenic activity of estrone
B. Relative estrogenic activity of β-estradiol
Ref. 1 - p. 945

1697. A. Relative estrogenic activity of estrone
B. Relative estrogenic activity of progesterone
Ref. 1 - pp. 949, 950

1698. A. Sodium retention produced by aldosterone
B. Sodium retention produced by cortisol
Ref. 1 - p. 962

ANSWER THE FOLLOWING QUESTIONS BY USING THE KEY OUTLINED BELOW:
A. If 1, 2 and 3 are correct
B. If 1 and 3 are correct
C. If 2 and 4 are correct
D. If only 4 is correct
E. If all four are correct

1699. 1. Alloxan causes selective destruction of alpha-cells of pancreas
2. Insulin loses physiological activity when treated with iodine
3. Formation of disulfide bonds in insulin molecule leads to loss of biological activity
4. Circulating insulin is destroyed by insulinase in the liver
Ref. 1 - pp. 973-976

1700. PROGESTERONE:
1. Differs from deoxycorticosterone at C-19
2. Can be synthesized by corpus luteum and adrenals
3. Is excreted in urine largely as pregnanediol glucosiduronate
4. Is relatively inactive when given orally
Ref. 1 - pp. 950, 958

1701. 1. The thyroid gland maintains a concentration of iodide higher than that found in extracellular fluid
2. The thyroid gland can oxidize iodine to iodide
3. 3, 3'-diiodothyronine is formed by condensation of two molecules of 3-monoiodotyrosine
4. Thyroxine is a constituent of various proteins
Ref. 1 - pp. 923-926

1702. CONCERNING CYCLIC AMP:
1. Adrenaline or nonadrenaline activate adenyl cyclase at β-receptor sites
2. Cyclic AMP is the "second messenger" in mediating cellular response to a hormone
3. Hormones which cause a decrease of cyclic AMP levels, are insulin and the prostaglandins
4. Caffeine inhibits β-adrenergic effects because it stimulates the enzyme which catalyzes the hydrolysis of cyclic AMP

Ref. 2 - pp. 455-456
Ref. 5 - pp. 554-555

1703. PRESENT CONCEPTS REGARDING THE MECHANISM OF HORMONE ACTION:
1. At present it seems probable that hormones can act via three different mechanisms namely stimulation of mRNA synthesis, stimulation of adenyl cyclase, alteration of membrane transport
2. All of the metabolic effects of the catecholamines can be accounted for by changes in the concentration of cyclic AMP in the affected cells
3. Glucagon stimulates adenyl cyclase in liver and adipose tissue, but not in skeletal muscle
4. A phosphodiesterase present in practically all tissues inactivates cyclic AMP. This enzyme is inhibited by citrate and pyrophosphate

Ref. 2 - p. 456
Ref. 5 - pp. 553-557

1704. IN DIABETES:
1. Glucose-6-phosphatase activity increases
2. Glucose-6-phosphate dehydrogenase activity increases
3. Glucose utilization is impaired
4. Gluconeogenesis is impaired

Ref. 1 - pp. 443, 448

1705.
1. The liver plays a minor role in regulating the blood level of thyroid hormone
2. Most of the iodine in the thyroid gland is in the form of iodide
3. In plasma the inorganic iodide concentration is greater than the protein-bound iodine concentration
4. The oxidation of I^- to organically bound iodine in the thyroid involves H_2O_2 and iodide peroxidase

Ref. 1 - p. 924

1706. IN ADRENAL CORTICAL INSUFFICIENCY:
1. Insulin sensitivity decreases
2. Noxious stimuli are more easily fatal
3. Steroid therapy is useless
4. Intestinal absorption of carbohydrate is impaired

Ref. 1 - pp. 963, 967

MATCH THE FOLLOWING HORMONES AND THEIR ACTION:

A. Prostaglandin
B. TSH
C. Somatotrophin
D. ACTH
E. FSH

1707. ___ Regulates secretion of corticosteroids
1708. ___ Controls formation and secretion of thyroid hormone
1709. ___ Affects estrogen secretion
1710. ___ Acts on smooth muscle and affects blood pressure
1711. ___ Growth hormone

Ref. 1 - pp. 918-920

MATCH THE HORMONES AND THEIR ACTION:

A. Melatonin
B. MSH
C. Thymosin
D. Calcitonin
E. Aldosterone

1712. ___ Pigment dispersal
1713. ___ Electrolyte metabolism
1714. ___ Pigment aggregation
1715. ___ Metabolism of Ca and P
1716. ___ Stimulation of lymphocytopoiesis Ref. 1 - pp. 918-920

MATCH THE HORMONES AND THEIR ACTION:

A. Secretin
B. Gastrin
C. Relaxin
D. Cholecystokinin
E. Pancreozymin

1717. ___ Contraction of gall bladder
1718. ___ Effect on muscle tone
1719. ___ Secretion of digestive enzymes
1720. ___ Secretion of stomach acid
1721. ___ Secretion of pancreatic juice Ref. 1 - pp. 919-920

RELATE THE ENDOCRINE GLANDS LISTED BELOW TO THEIR PRINCIPAL SITES OF ACTION:

A. Thymosin
B. ACTH
C. Calcitonin
D. Epinephrine
E. Progesterone

1722. ___ Adrenal cortex
1723. ___ Lymphoid tissue
1724. ___ Skeleton
1725. ___ Uterus
1726. ___ Adipose tissue Ref. 1 - pp. 918-920

EACH OF THE QUESTIONS OR INCOMPLETE STATEMENTS BELOW IS FOLLOWED BY 5 SUGGESTED ANSWERS OR COMPLETIONS. SELECT THE ONE WHICH IS BEST IN EACH CASE:

1727. INGESTION OF BENZOIC ACID IN MAN RESULTS IN AN INCREASE IN URINARY EXCRETION OF:
A. Phenylalanine
B. Norepinephrine
C. Phenylacetic acid
D. Hippuric acid
E. Benzoyl sulfate
Ref. 2 - pp. 733-734

1728. THE SYNTHESIS OF HIPPURIC ACID INVOLVES A(N):
A. Transglycosylase
B. Amidase
C. Deaminase
D. Aminidase
E. Phosphorylase
Ref. 2 - pp. 733-734

1729. MANY SPECIES OF BACTERIA REQUIRE p-AMINOBENZOIC ACID IN THE SYNTHESIS OF:
A. Sulfanilamide
B. Pantothenic acid
C. Hippuric acid
D. Vitamin K
E. Folic acid
Ref. 2 - p. 832

1730. DICUMAROL IS AN ANTIMETABOLITE OF:
A. Vitamin K
B. Heparin
C. Prothrombin
D. Vitamin B_{12}
E. Folic acid
Ref. 1 - p. 728

1731. SMALL AMOUNTS OF OXALIC ACID MAY BE EXCRETED IN THE URINE. THIS ACID CAN ARISE FROM:
A. Dietary oxalic acid
B. Ascorbic acid
C. Carbohydrate
D. Deoxypyridoxine
E. All of these
Ref. 1 - pp. 1030-1044

1732. THE BIOSYNTHESIS OF CREATINE INVOLVES 3 AMINO ACIDS. THESE ARE:
A. Methionine, glycine and histidine
B. Methionine, glycine and arginine
C. Methionine, lysine and arginine
D. Methionine, glycine and cystine
E. Glycine, arginine and lysine
Ref. 2 - pp. 396-398

1733. AN ENZYME CLOSELY ASSOCIATED WITH ACTOMYOSIN IS:
A. ATPase
B. Adenyl cyclase
C. Translocase
D. Relaxing factor
E. Hexokinase
Ref. 2 - pp. 562-564

1734. WHICH OF THE FOLLOWING STATEMENTS CONCERNING THE PRESENCE OF HYDROXYPROLINE IN COLLAGEN IS MOST NEARLY CORRECT?:
A. Proline is first incorporated into the peptide chain. Some of the prolines are then hydroxylated enzymatically
B. Hydroxyproline is incorporated into the peptide chain by way of hypro-tRNA
C. Hydroxyproline is first incorporated into the peptide chain and is then partially dehydroxylated
D. Proline, but not hydroxyproline is incorporated into the peptide chain. Some of the prolines are then hydroxylated spontaneously in the presence of oxygen and ferric ion
E. In scurvy the production of hydroxyproline from proline is greatly accelerated
Ref. 2 - p. 545

EACH GROUP OF QUESTIONS BELOW CONSISTS OF 5 LETTERED HEADINGS FOLLOWED BY A LIST OF NUMBERED WORDS OR PHRASES. FOR EACH NUMBERED WORD OR PHRASE SELECT THE ONE LETTERED HEADING THAT IS MOST CLOSELY RELATED TO IT:

MATCH THE PROPER ANTIBIOTIC WITH THE MOST CORRECT STATEMENT CONCERNING IT:

A. Pyridoxine antagonist
B. Inhibitor of protein synthesis
C. Interferes with biosynthesis of cell wall substance in gram-positive bacteria
D. Inhibitor of oxidative phosphorylation
E. Inhibits DNA-dependent RNA polymerase

1735. ___ Penicillin
1736. ___ Gramicidin
1737. ___ Isoniazide
1738. ___ Chloramphenicol
1739. ___ Actinomycin D

Ref. 1 - pp. 670, 912
Ref. 3 - pp. 551-552, 686, 954

A. Indican
B. Glucuronic acid
C. Puromycin
D. Thiouracil

1740. ___ "Detoxication" product of indole found in urine
1741. ___ "Detoxication" of many alcohols and phenols requires ___________
1742. ___ Interferes with protein synthesis
1743. ___ Inhibits formation of thyroxine

Ref. 1 - pp. 670, 843, 930

A. p-Mercuribenzoate
B. Methylcholanthrene
C. Digitonin
D. Rhodanese
E. Cryptoxanthin

1744. ___ Carcinogen
1745. ___ Steroid precipitant
1746. ___ Inhibits enzymes by combining with sulfhydryl groups
1747. ___ Provitamin A
1748. ___ Detoxifies cyanide

Ref. 1 - pp. 79, 83, 239, 598, 1048

A. Caffeine
B. 8-Azaguanine
C. Bromobenzene
D. Pyrithiamine
E. Thiochrome

1749. ___ Antagonist of thiamine
1750. ___ Oxidation product of thiamine
1751. ___ Excreted as a mercapturic acid
1752. ___ Metabolic antagonist of adenine
1753. ___ 1, 3, 7-trimethylxanthine

Ref. 1 - pp. 184, 580, 682, 1020

A. Amytal
B. Cyanide
C. Rotenone
D. Antimycin A
E. 2-Pyridine aldoxime methiodide

1754. ___ Inhibits NADH dehydrogenase
1755. ___ Inhibit(s) reduction of cytochromes a and a_3
1756. ___ Inhibit(s) reduction of cytochromes c_1, c and a_3
1757. ___ Inhibit(s) reduction of ubiquinone
1758. ___ Antidote to "nerve gases" Ref. 1 - pp. 339, 886

A. Notatin
B. Monoamine oxidase inhibitor
C. 8-Azaguanine
D. Acetazolamide

1759. ___ Glucose oxidase
1760. ___ Inhibits serotonin metabolism
1761. ___ Inhibitor of carbonic anhydrase Ref. 1 - pp. 682, 767, 864
1762. ___ Interferes with protein synthesis Ref. 3 - p. 541

A. Nicotine
B. AT-10 (Dihydrotachysterol)
C. Ethionine
D. Phlorhizin
E. Ergothioneine

1763. ___ A betaine
1764. ___ Depresses renal threshold for glucose
1765. ___ Combines with ferroprotoporphyrin
1766. ___ Used in treatment of rickets Ref. 1 - pp. 170, 612, 837,
1767. ___ Produces fatty liver 1008, 1052

ANSWER THE FOLLOWING QUESTIONS BY USING THE KEY OUTLINED BELOW:
A. If 1, 2 and 3 are correct
B. If 1 and 3 are correct
C. If 2 and 4 are correct
D. If only 4 is correct
E. If all four are correct

1768. p-AMINOHIPPURIC ACID:
1. A peptide
2. Is synthesized primarily in the kidneys
3. Used in clearance studies Ref. 1 - pp. 579, 819-820
4. Antagonist of hippuric acid Ref. 2 - p. 757

1769. ACETAZOLAMIDE IS A SULFONAMIDE COMPOUND WHICH:
1. Inhibits growth of gram-negative bacteria
2. Promotes ion transport
3. Is a potent folic acid antagonist
4. Inhibits carbonic anhydrase Ref. 1 - p. 767

1770. AMINOPTERIN:
1. Is the 4-amino analogue of pteroylgutamic acid
2. Inhibits reduction of folic acid to the tetrahydro derivative
3. Inhibits synthesis of thymidylic acid
4. Antagonizes the action of 9-methylaminopterin
Ref. 1 - p. 1035

EACH OF THE QUESTIONS OR INCOMPLETE STATEMENTS BELOW IS FOLLOWED BY 5 SUGGESTED ANSWERS OR COMPLETIONS. SELECT THE ONE WHICH IS BEST IN EACH CASE:

1771. IRON IS STORED IN THE LIVER ALMOST ENTIRELY IN THE FORM OF:
A. Ferric ion
B. Hemosiderin
C. Transferrin
D. Ferritin
E. Ferrous ion
Ref. 1 - p. 749

1772. THE VISUAL IMPULSE IS ASSOCIATED WITH THE:
A. Transformation of rhodopsin to retinol
B. The reduction of TPN
C. The condensation of opsin with vitamin C aldehyde
D. The hydrolysis of visual purple
E. All of the above
Ref. 1 - pp. 901-903

1773. A RISE OF INORGANIC CALCIUM AND PHOSPHATE IN THE BLOOD IS INDICATIVE OF:
A. Vitamin C deficiency
B. Vitamin D deficiency
C. Hypervitaminosis D
D. Insulin deficiency
E. Low serum protein
Ref. 1 - p. 894

1774. THE CHIEF PIGMENT OF SKIN IS:
A. Dihydroxyphenylalanine
B. Hyaluronic acid
C. Melanin
D. Dopamine
E. Melatonin
Ref. 1 - p. 608

1775. COLLAGEN IS A MAJOR COMPONENT OF:
A. Blood
B. Bone
C. Liver
D. Heparin
E. Bile
Ref. 1 - p. 890

1776. THE PREDOMINANT MUCOPOLYSACCHARIDE OF CARTILAGE IS:
A. Elastin
B. Collagen
C. Hyaluronic acid
D. Chondroitin sulfate
E. Heparin
Ref. 1 - p. 878

1777. COLLAGEN IS UNIQUE IN THAT IT CONTAINS MUCH:
A. Peptone
B. Hydroxyproline
C. Cystine
D. Methionine
E. Hyaluronidase
Ref. 1 - pp. 116; 872

1778. THE METABOLISM OF THE BRAIN:
A. Is mainly dependent upon its large glycogen stores
B. Is independent of the blood glucose level
C. Accounts for 25% of the oxygen consumption of the body at rest
D. Is influenced greatly by the uptake of fatty acids from the blood
E. None of these apply
Ref. 1 - p. 862

1779. THE ETIOLOGIC AGENT IN OSTEOGENIC LATHYRISM IS:
A. Lysine deficiency
B. Unknown
C. Gamma-aminobutyrate
D. Hydroxyproline
E. Aminopropionitrile
Ref. 1 - pp. 874, 877

ANSWER THE FOLLOWING GROUP OF QUESTIONS BY USING THE KEY OUTLINED BELOW:

A. If the item is associated with A only
B. If the item is associated with B only
C. If the item is associated with both A and B
D. If the item is associated with neither A nor B

A. Liver mitochondria
B. Liver microsomes
C. Both
D. Neither

1780. ___ Protein synthesis
1781. ___ Phosphatases
1782. ___ Glycolytic system
1783. ___ Krebs cycle
1784. ___ Electron transport Ref. 1 - pp. 286, 372

EACH GROUP OF QUESTIONS BELOW CONSISTS OF 5 LETTERED HEADINGS FOLLOWED BY A LIST OF NUMBERED WORDS OR PHRASES. FOR EACH NUMBERED WORD OR PHRASE SELECT THE ONE LETTERED HEADING THAT IS MOST CLOSELY RELATED TO IT:

A. RNA of nerve cells
B. Serotonin
C. Diisopropylfluorophosphate
D. Gamma-aminobutyric acid
E. Norepinephrine

1785. ___ Formed in brain by alpha-decarboxylation of glutamic acid
1786. ___ Synthesized from tryptophan in brain
1787. ___ Nissl substance
1788. ___ Irreversible inhibitor of acetylcholine esterase
1789. ___ Secreted by postganglionic nerve fibers
Ref. 1 - pp. 862-864, 867, 869

A. Collagen
B. Chondroitin sulfate C
C. Hyaluronic acid
D. Dermatan sulfate
E. Gelatin

1790. ___ A polymer of glucuronic acid and acetylgalactosamine sulfate
1791. ___ Results from action of boiling water on collagen
1792. ___ Most abundant protein in human body
1793. ___ Present in high concentration in Wharton's jelly
1794. ___ Contains iduronic acid Ref. 1 - pp. 52-53, 871-872

CHAPTER VII - SPECIAL TOPICS : METABOLIC ANTAGONISTS, SPECIALIZED TISSUES

ANSWER THE FOLLOWING QUESTIONS BY USING THE KEY OUTLINED BELOW:

A. If A is greater than B
B. If B is greater than A
C. If A and B are equal or nearly equal

1795. A. Isocitric dehydrogenase in bone
B. Isocitric dehydrogenase in liver — Ref. 1 - pp. 330, 893-894

1796. A. Turnover of phosphate in tooth dentine
B. Turnover of phosphate in tooth enamel
Ref. 1 - p. 896

1797. A. Turnover of phosphate in dentine of teeth
B. Turnover of phosphate in long bones
Ref. 1 - p. 896

1798. A. Mineral content of dentine
B. Mineral content of enamel — Ref. 1 - p. 896

1799. A. ATP content of actin
B. ATP content of myosin
Ref. 1 - pp. 849-850

1800. A. ATP-ase activity of actin
B. ATP-ase activity of myosin
Ref. 1 - pp. 849-850

REFERENCES

1. White, A., P. Handler and E. L. Smith: Principles of Biochemistry, Fourth Edition, 1968, The Blakiston Division, McGraw-Hill Book Co., Inc., New York, N. Y.

2. Orten, J. M. and O. W. Neuhaus: Biochemistry, Eighth Edition, 1970, The C. V. Mosby Co., St. Louis, Mo.

3. Mahler, H. R. and E. H. Cordes: Biological Chemistry, Second Edition, 1971, Harper and Row, New York, N. Y.

4. McElroy, W. D.: Cell Physiology and Biochemistry, Third Edition, 1971, Prentice-Hall, Inc., Englewood Cliffs, N. J.

5. McGilvery, R. W.: Biochemistry, W. B. Saunders Co., Philadelphia, Penn., 1970

The author has taken great pains to check thoroughly the questions and answers. However, in a volume of this size, some ambiguities and possible inaccuracies may appear. Therefore, it in doubt, consult your references.

THE PUBLISHERS

CHAPTER I

1. A
2. C
3. E
4. D
5. B
6. D
7. C
8. C
9. D
10. B
11. E
12. A
13. D
14. B
15. D
16. B
17. D
18. E
19. A
20. C
21. D
22. D
23. C
24. D
25. A
26. A
27. C
28. D
29. E
30. A
31. B
32. C
33. D
34. E
35. A
36. C
37. B
38. D
39. B
40. C
41. E
42. A
43. D
44. A
45. E
46. C
47. B
48. B
49. E
50. C
51. A
52. D
53. B
54. D
55. A
56. E
57. C
58. A
59. D
60. E
61. B
62. C
63. E
64. D
65. C
66. B
67. A
68. C
69. B
70. A
71. D
72. E
73. E
74. A
75. D
76. B
77. C
78. C
79. B
80. C
81. A
82. C
83. D
84. B
85. A
86. C
87. A
88. A
89. C
90. B
91. C
92. E
93. D
94. E
95. C
96. A
97. B
98. E
99. D
100. C
101. B
102. A
103. A
104. B
105. C
106. D
107. E
108. C
109. B
110. E
111. A
112. D
113. C
114. E
115. D
116. A
117. B
118. C
119. D
120. E
121. A
122. B
123. E
124. A
125. C
126. D
127. B
128. E
129. C
130. B
131. C
132. A
133. B
134. D
135. E
136. C
137. B
138. B
139. A
140. A
141. B
142. A
143. B
144. C
145. A
146. B
147. A
148. B
149. A
150. C
151. B
152. B
153. B
154. D
155. C
156. C
157. D
158. E
159. E
160. C
161. B
162. D
163. A

CHAPTER II

164. A
165. C
166. A
167. C
168. D
169. B
170. A
171. A
172. D
173. D
174. D
175. C
176. E
177. B
178. A
179. B
180. C
181. B
182. A
183. C
184. B
185. E
186. D
187. A
188. C
189. A
190. B
191. E
192. C
193. C
194. D
195. B
196. C
197. E
198. D
199. D
200. D
201. A
202. E
203. E
204. E
205. C
206. C
207. A
208. C
209. D
210. C
211. E
212. D
213. E
214. C
215. B
216. A
217. C
218. D
219. A
220. C
221. B
222. E
223. E
224. C
225. D
226. A
227. B
228. A
229. B
230. C
231. E
232. D
233. A
234. C
235. E
236. D
237. B
238. A
239. E
240. B
241. D
242. C
243. C
244. B
245. A
246. D
247. E
248. D
249. C
250. E
251. B
252. A
253. C
254. B
255. E
256. A
257. D
258. C
259. B
260. D
261. D
262. D
263. A
264. B
265. A
266. C
267. D
268. B
269. C
270. A
271. A
272. B
273. B
274. C
275. B
276. C
277. A
278. B
279. C
280. A
281. B
282. A
283. B
284. D
285. A
286. B
287. A
288. C
289. D
290. C
291. C
292. D
293. B
294. C
295. A
296. B
297. A

298. C
299. B
300. B
301. B
302. C
303. B
304. C
305. B
306. A
307. B
308. B
309. A
310. A
311. B
312. C
313. A
314. C
315. B
316. A
317. B
318. B
319. C
320. A
321. A
322. B
323. A
324. D
325. A
326. B
327. A
328. B
329. A
330. A
331. E
332. C
333. B
334. E
335. A
336. D
337. A
338. B
339. D
340. E
341. C
342. B
343. C
344. G
345. E
346. F
347. D
348. A
349. B
350. D
351. C
352. E
353. E
354. A
355. C
356. A
357. D
358. C
359. B
360. E
361. B
362. D
363. D
364. D
365. C
366. E
367. B
368. E
369. A
370. C
371. D
372. D
373. B
374. A
375. C
376. E
377. C
378. A
379. E
380. B
381. D
382. D
383. B
384. E
385. C
386. A
387. A
388. D
389. B
390. E
391. C
392. B
393. E
394. D
395. A
396. C
397. B
398. A
399. A
400. C
401. A
402. B
403. A
404. D
405. C
406. A
407. B
408. A
409. D
410. B
411. E
412. D
413. A
414. C
415. C
416. E
417. A
418. D
419. B
420. D
421. B
422. A
423. C
424. E
425. A
426. A
427. E
428. C
429. D
430. E
431. B
432. D
433. A
434. C
435. D
436. E
437. B
438. C
439. A
440. D
441. E
442. C
443. D
444. C
445. C
446. A
447. C
448. A
449. D
450. C
451. B
452. D
453. B
454. D
455. E
456. E
457. B
458. C
459. C
460. D
461. B
462. B
463. A
464. E
465. D
466. C
467. E
468. A
469. D
470. B
471. D
472. A
473. E
474. C
475. B
476. E
477. D
478. A
479. C
480. B
481. C
482. A
483. B
484. E
485. D
486. A
487. A
488. B
489. A
490. B
491. B
492. B
493. C
494. B
495. A
496. B
497. A
498. B
499. A
500. B
501. A
502. A
503. C
504. C
505. E
506. A
507. A
508. E
509. A
510. D
511. C
512. E
513. B
514. B
515. B
516. B
517. A
518. D
519. B
520. D
521. E
522. D
523. B
524. A
525. C
526. D
527. B
528. B
529. E
530. B
531. D
532. C
533. B
534. C
535. E
536. D
537. C
538. B
539. B
540. A
541. A
542. B
543. C
544. C
545. A
546. B
547. A
548. B
549. A
550. C
551. E
552. B
553. A
554. C
555. D
556. C
557. E
558. B
559. D
560. A
561. D
562. B
563. A
564. B
565. E
566. C
567. D
568. A
569. E
570. C
571. C
572. C
573. A
574. D
575. E
576. D
577. E
578. A
579. D
580. E
581. C
582. B
583. D
584. E
585. B
586. A
587. C
588. A
589. C
590. B
591. E
592. D
593. C
594. A
595. B
596. E
597. D
598. A
599. B
600. C
601. E
602. D
603. D
604. B
605. E
606. D
607. D
608. B
609. D
610. C
611. B
612. A
613. E
614. B
615. A
616. D
617. C
618. E
619. C
620. E
621. B
622. D
623. A
624. B
625. D
626. C
627. E
628. A
629. D
630. B
631. A
632. C
633. E
634. A
635. B
636. D
637. C
638. E
639. A
640. C
641. B
642. E
643. D
644. C
645. D

646. A
647. E
648. B
649. E
650. B
651. A
652. C
653. D
654. B
655. C
656. A
657. E
658. D
659. B
660. C
661. A
662. E
663. D
664. C
665. A
666. B
667. D
668. E
669. A
670. B
671. B
672. B
673. B
674. B
675. A
676. C
677. B
678. B
679. B
680. A
681. B
682. B
683. B
684. A
685. D
686. E
687. A
688. B
689. E
690. A
691. C
692. E
693. A
694. D
695. B
696. A
697. B
698. E
699. C
700. D
701. A
702. B
703. D
704. C
705. E
706. A
707. D
708. B
709. E
710. C
711. B
712. B
713. E
714. A
715. C
716. C
717. C
718. E
719. B
720. D
721. B
722. B
723. B
724. B
725. C
726. D
727. C
728. B
729. B
730. D
731. A
732. D
733. D
734. C
735. C
736. A
737. E
738. B
739. B
740. C
741. B
742. C
743. B
744. C
745. A
746. C
747. D
748. C
749. A
750. A
751. D
752. C
753. C
754. D
755. D
756. B
757. D
758. B
759. C
760. E
761. A
762. A
763. B
764. C
765. C
766. C
767. A
768. C
769. B
770. C
771. B
772. C
773. B
774. A
775. B
776. B
777. A
778. A
779. C
780. A
781. B
782. A
783. A
784. B
785. A
786. C
787. B
788. B
789. E
790. D
791. C
792. A
793. E
794. A
795. C
796. B
797. D
798. C
799. A
800. B
801. D
802. E
803. B
804. E
805. A
806. D
807. C
808. E
809. C
810. B
811. A
812. D
813. B
814. A
815. D
816. C
817. E
818. A
819. B
820. E
821. C
822. D
823. E
824. D
825. C
826. B
827. A
828. E
829. D
830. C
831. B
832. A
833. A
834. B
835. C
836. E
837. D
838. D
839. C
840. E
841. B
842. A
843. B
844. C
845. E
846. D
847. A
848. B
849. A
850. C
851. C
852. A
853. C
854. A
855. B
856. C
857. D
858. E
859. A
860. B
861. C
862. B
863. E
864. C
865. A
866. D
867. A
868. B
869. A
870. D
871. D
872. E
873. C
874. B
875. D
876. E
877. A
878. C
879. B
880. E
881. D
882. C
883. A
884. B
885. A
886. C
887. B
888. C
889. D
890. A
891. E
892. C
893. B
894. D
895. E
896. C
897. B
898. A
899. B
900. E
901. C
902. A
903. D
904. C
905. B
906. D
907. A
908. E
909. A
910. C
911. E
912. B
913. D
914. B
915. C
916. A
917. B
918. A
919. A
920. A
921. C
922. B
923. B
924. D
925. A
926. D
927. D
928. B
929. A
930. B
931. A
932. C
933. E
934. A
935. E
936. D
937. E
938. C
939. B
940. D
941. A
942. D
943. E
944. C
945. B
946. B
947. B
948. C
949. A
950. B
951. C
952. E
953. B
954. C
955. B
956. E
957. C
958. B
959. C
960. A
961. B
962. B
963. A
964. A
965. B
966. B
967. A
968. C
969. D
970. A
971. A
972. C
973. A
974. D
975. C
976. C
977. E
978. C
979. B
980. A
981. D
982. A
983. D
984. B
985. E
986. C
987. E
988. D
989. B
990. A
991. C
992. A
993. C

994. E
995. B
996. D
997. D
998. C
999. A
1000. B
1001. E
1002. E
1003. B
1004. C
1005. E
1006. A
1007. D
1008. C
1009. B
1010. E
1011. A
1012. D
1013. B
1014. E
1015. C
1016. A
1017. D
1018. B
1019. C
1020. D
1021. E
1022. A
1023. E
1024. A
1025. B
1026. C
1027. D
1028. B
1029. E
1030. D
1031. A
1032. C
1033. C
1034. A
1035. A
1036. C
1037. A
1038. B
1039. C
1040. A
1041. B
1042. D
1043. B
1044. D
1045. B
1046. B
1047. C
1048. B
1049. C
1050. A
1051. C
1052. A
1053. C
1054. C
1055. B
1056. A
1057. B
1058. C
1059. A
1060. C
1061. B
1062. A
1063. D
1064. B
1065. B

CHAPTER III

1066. D
1067. B
1068. B
1069. B
1070. E
1071. A
1072. C
1073. B
1074. E
1075. E
1076. A
1077. B
1078. C
1079. E
1080. E
1081. C
1082. A
1083. D
1084. C
1085. C
1086. D
1087. A
1088. C
1089. A
1090. B
1091. B
1092. D
1093. B
1094. A
1095. B
1096. D
1097. A
1098. B
1099. B
1100. D
1101. C
1102. D
1103. C
1104. A
1105. B
1106. C
1107. B
1108. B
1109. D
1110. A
1111. E
1112. C
1113. B
1114. D
1115. E
1116. C
1117. A
1118. D
1119. A
1120. C
1121. E
1122. B
1123. D
1124. B
1125. A
1126. E
1127. C
1128. B
1129. E
1130. A
1131. D
1132. C
1133. B
1134. D
1135. A
1136. E
1137. C
1138. C
1139. E
1140. A
1141. B
1142. D
1143. D
1144. B
1145. C
1146. E
1147. A
1148. C
1149. D
1150. E
1151. B
1152. A
1153. A
1154. D
1155. E
1156. B
1157. C
1158. C
1159. E
1160. A
1161. D
1162. B
1163. C
1164. B
1165. A
1166. C
1167. D
1168. D
1169. B
1170. A
1171. D

CHAPTER IV

1172. D
1173. B
1174. A
1175. C
1176. E
1177. A
1178. E
1179. C
1180. B
1181. D
1182. A
1183. D
1184. C
1185. E
1186. C
1187. E
1188. D
1189. E
1190. D
1191. C
1192. E
1193. A
1194. A
1195. C
1196. A
1197. B
1198. C
1199. A
1200. B
1201. B
1202. B
1203. E
1204. C
1205. C
1206. E
1207. C
1208. D
1209. D
1210. C
1211. A
1212. C
1213. A
1214. B
1215. A
1216. D
1217. A
1218. E
1219. C
1220. B
1221. C
1222. D
1223. B
1224. B
1225. E
1226. A
1227. A
1228. B
1229. E
1230. C
1231. D
1232. B
1233. A
1234. B
1235. B
1236. B
1237. D
1238. C
1239. A
1240. B
1241. C
1242. D
1243. B
1244. B
1245. B
1246. C
1247. B
1248. D
1249. A
1250. B
1251. D
1252. B
1253. A
1254. A
1255. C
1256. D
1257. A
1258. A
1259. B
1260. C
1261. D
1262. D
1263. B
1264. B
1265. D
1266. A
1267. D
1268. C
1269. B
1270. D
1271. E
1272. A
1273. C
1274. A
1275. C
1276. B
1277. E
1278. D
1279. D
1280. D
1281. B
1282. A
1283. D
1284. C
1285. C
1286. D
1287. C
1288. B
1289. A
1290. E
1291. A
1292. D
1293. B
1294. A
1295. A
1296. B
1297. A
1298. A
1299. A
1300. A
1301. A
1302. B
1303. C
1304. B
1305. C
1306. A
1307. B
1308. A
1309. C
1310. D
1311. A
1312. E
1313. B
1314. E
1315. C
1316. E
1317. E
1318. B
1319. C
1320. B
1321. A
1322. C
1323. C
1324. D
1325. E
1326. E
1327. B
1328. A
1329. C
1330. D
1331. E
1332. A
1333. C
1334. B
1335. E

1336. D
1337. B
1338. D
1339. C
1340. A
1341. E
1342. D
1343. B
1344. C
1345. A
1346. E
1347. A
1348. D
1349. E
1350. C
1351. B
1352. E
1353. B
1354. A
1355. C
1356. D
1357. B
1358. C
1359. D
1360. A
1361. E
1362. A
1363. B
1364. C
1365. D
1366. B
1367. C
1368. B
1369. A
1370. B
1371. D
1372. A
1373. C
1374. B
1375. A
1376. B
1377. A
1378. B
1379. A
1380. C
1381. B
1382. B
1383. A
1384. B
1385. B
1386. E
1387. D
1388. C

CHAPTER V

1389. A
1390. D
1391. C
1392. E
1393. B
1394. D
1395. C
1396. E
1397. A
1398. D
1399. C
1400. E
1401. D
1402. A
1403. A
1404. B
1405. B
1406. D
1407. C
1408. E
1409. A
1410. C
1411. E
1412. D
1413. A
1414. B
1415. C
1416. B
1417. E
1418. A
1419. D
1420. D
1421. B
1422. A
1423. C
1424. E
1425. D
1426. B
1427. A
1428. E
1429. C
1430. A
1431. C
1432. D
1433. D
1434. B
1435. C
1436. E
1437. D
1438. A
1439. B
1440. B
1441. A
1442. C
1443. A
1444. B
1445. E
1446. C
1447. A
1448. B
1449. D
1450. C
1451. C
1452. D
1453. A
1454. D
1455. B
1456. A
1457. B
1458. B
1459. B
1460. B
1461. C
1462. A
1463. B
1464. B
1465. A
1466. A
1467. B
1468. C
1469. D
1470. A
1471. B
1472. C
1473. B
1474. B
1475. A
1476. A
1477. B
1478. B
1479. B
1480. A
1481. B
1482. C
1483. A
1484. D
1485. B
1486. B
1487. C
1488. C
1489. A
1490. A
1491. E
1492. C
1493. B
1494. D
1495. B
1496. E
1497. A
1498. D
1499. C
1500. A
1501. C
1502. E
1503. B
1504. D
1505. F
1506. C
1507. B
1508. D
1509. A
1510. A
1511. B
1512. B
1513. E
1514. B
1515. D
1516. A
1517. D
1518. E
1519. B
1520. A
1521. C
1522. C
1523. C
1524. B
1525. D
1526. D
1527. B
1528. A
1529. B
1530. C
1531. D
1532. D
1533. E
1534. C
1535. C
1536. D
1537. B
1538. D
1539. C
1540. D
1541. A
1542. E
1543. C
1544. B
1545. B
1546. A
1547. D
1548. E
1549. C
1550. E
1551. B
1552. A
1553. C
1554. D
1555. D
1556. A
1557. B
1558. C
1559. E
1560. B
1561. A
1562. C
1563. A
1564. B
1565. B
1566. D
1567. D
1568. C
1569. E
1570. D
1571. A
1572. B
1573. A
1574. B
1575. B
1576. C
1577. A
1578. B

CHAPTER VI

1579. A
1580. E
1581. B
1582. D
1583. C
1584. E
1585. C
1586. D
1587. A
1588. B
1589. A
1590. B
1591. E
1592. D
1593. C
1594. C
1595. C
1596. C
1597. C
1598. A
1599. A
1600. D
1601. A
1602. E
1603. A
1604. E
1605. E
1606. B
1607. B
1608. C
1609. D
1610. B
1611. B
1612. C
1613. B
1614. C
1615. D
1616. E
1617. A
1618. C
1619. A
1620. D
1621. D
1622. A
1623. C
1624. E
1625. B
1626. E
1627. A
1628. C
1629. D
1630. B
1631. C
1632. B
1633. A
1634. A
1635. B
1636. B
1637. E
1638. D
1639. A
1640. C
1641. C
1642. A
1643. B
1644. E
1645. D
1646. A
1647. C
1648. E
1649. B
1650. D
1651. C
1652. D
1653. E
1654. A
1655. B
1656. A
1657. C
1658. B
1659. A
1660. B
1661. A
1662. D
1663. A
1664. B
1665. B
1666. B
1667. C
1668. B
1669. A
1670. D
1671. D
1672. C
1673. A
1674. B
1675. C
1676. C
1677. D

1678. D
1679. C
1680. B
1681. A
1682. B
1683. A
1684. A
1685. A
1686. B
1687. B
1688. C
1689. C
1690. A
1691. A
1692. B
1693. A
1694. A
1695. A
1696. B
1697. A
1698. A
1699. C
1700. A
1701. B
1702. A
1703. E
1704. B
1705. D
1706. C
1707. D
1708. B
1709. E
1710. A
1711. C
1712. B
1713. E
1714. A
1715. D
1716. C
1717. D
1718. C
1719. E
1720. B
1721. A
1722. B
1723. A
1724. C
1725. E
1726. D

CHAPTER VII

1727. D
1728. B
1729. E
1730. A
1731. E
1732. B
1733. A
1734. A
1735. C
1736. D
1737. A
1738. B
1739. E
1740. A
1741. B
1742. C
1743. D
1744. B
1745. C
1746. A
1747. E
1748. D
1749. D
1750. E
1751. C
1752. B
1753. A
1754. C
1755. B
1756. D
1757. A
1758. E
1759. A
1760. B
1761. D
1762. C
1763. E
1764. D
1765. A
1766. B
1767. C
1768. B
1769. D
1770. A
1771. D
1772. A
1773. C
1774. C
1775. B
1776. D
1777. B
1778. C
1779. E
1780. B
1781. B
1782. D
1783. A
1784. C
1785. D
1786. B
1787. A
1788. C
1789. E
1790. B
1791. E
1792. A
1793. C
1794. D
1795. B
1796. A
1797. B
1798. A
1799. A
1800. B

OTHER BOOKS AVAILABLE

Quan.	ITEMS	Code	Unit Price
	MEDICAL EXAM REVIEW BOOKS		
	Vol. 1 Comprehensive	101	$ 12.00
	Vol. 2 Clinical Medicine	102	7.00
	Vol. 2A Txtbk. Stdy. Guide Int. Med.	123	7.00
	Vol. 2B Txtbk. Stdy. Guide Int. Med.	130	7.00
	Vol. 3 Basic Sciences	103	7.00
	Vol. 4 Obstetrics-Gynecology	104	7.00
	Vol. 5 Surgery	105	7.00
	Vol. 5A Textbk. Stdy. Guide Surgery	150	7.00
	Vol. 6 Public Health & Prev. Medicine	106	7.00
	Vol. 8 Psychiatry & Neurology	108	7.00
	Vol. 11 Pediatrics	111	7.00
	Vol. 12 Anesthesiology	112	7.00
	Vol. 13 Orthopaedics	113	10.00
	Vol. 14 Urology	114	10.00
	Vol. 15 Ophthalmology	115	10.00
	Vol. 16 Otolaryngology	116	10.00
	Vol. 17 Radiology	117	10.00
	Vol. 18 Thoracic Surgery	118	10.00
	Vol. 19 Neurological Surgery	119	15.00
	Vol. 20 Physical Medicine	128	10.00
	Vol. 21 Dermatology	127	10.00
	Vol. 22 Gastroenterology	141	10.00
	Vol. 23 Child Psychiatry	126	10.00
	ECFMG Exam Review - Part One	120	7.00
	ECFMG Exam Review - Part Two	121	7.00
	BASIC SCIENCE REVIEW BOOKS		
	Anatomy Review	201	6.00
	Biochemistry Review	202	6.00
	Microbiology Review	203	6.00
	Pathology Review	204	6.00
	Pharmacology Review	205	6.00
	Physiology Review	206	6.00
	Heart & Vascular Syst. Basic Sciences	212	7.00
	SPECIALTY BOARD REVIEW BOOKS		
	Pediatrics Specialty Board Review	301	10.00
	Surgery Specialty Board Review	302	10.00
	Int. Med. Specialty Board Review	303	10.00
	Obst.-Gyn. Specialty Board Review	304	10.00
	Pathology Specialty Board Review	305	10.00

Quan.	ITEMS	Code	Unit Price
	SPEC. BOARD REV. BKS. *(Cont'd.)*		
	Neurology Specialty Board Review	306	10.00
	Psychiatry Specialty Board Review	307	10.00
	Physical Med. Specialty Board Review	308	10.00
	STATE BOARD REVIEW BOOKS		
	Medical State Board Exam. Rev. - Part 1	411	9.00
	Medical State Board Exam. Rev. - Part 2	412	9.00
	Pharmacy Exam. Review Book - Vol. 1	421	7.00
	Dental Exam. Review Book - Vol. 1	431	7.50
	Dental Exam. Review Book - Vol. 2	432	7.50
	Dental Exam. Review Book - Vol. 3	433	7.50
	X-Ray Technology Exam. Rev. - Vol. 1	441	7.00
	X-Ray Technology Exam. Rev. - Vol. 2	442	7.00
	Medical Techn. Exam. Rev. - Vol. 1	451	7.00
	Medical Techn. Exam. Rev. - Vol. 2	452	7.00
	Dental Hygiene Exam. Rev. - Vol. 1	461	7.00
	Emergency Med. Techn. Exam. - Vol. 1	465	7.00
	Emergency Med. Techn. Exam. - Vol. 2	466	7.00
	Inhalation Therapy Exam. Rev. - Vol. 1	471	7.00
	Inhalation Therapy Exam. Rev. - Vol. 2	344	7.00
	Physical Therapy Exam. Rev. - Vol. 1	481	7.00
	Physical Therapy Exam. Rev. - Vol. 2	482	7.00
	Medical Record Library Science - Vol. 1	496	7.00
	Occupational Therapy Exam. - Vol. 1	475	7.00
	Medical Librarian Exam. - Vol. 1	495	7.00
	Immunology Exam. Review Book - Vol. 1	424	7.00
	Laboratory Asst. Exam. Rev. Bk. - Vol. 1	455	7.00
	Cardio-Pulmonary Techn. Exam. Rev. - Vol. 1	473	7.00
	NURSING EXAM REVIEW BOOKS		
	Vol. 1 Medical Surgical Nursing	501	4.00
	Vol. 2 Psychiatric Nursing	502	4.00
	Vol. 3 Maternal-Child Health Nursing	503	4.00
	Vol. 4 Basic Sciences	504	4.00
	Vol. 5 Anatomy and Physiology	505	4.00
	Vol. 6 Pharmacology	506	4.00
	Vol. 7 Microbiology	507	4.00
	Vol. 8 Nutrition & Diet Therapy	508	4.00
	Vol. 9 Community Health	509	4.00
	Vol. 10 History & Law of Nursing	510	4.00

(Continued on reverse side)

Prices subject to change

OTHER BOOKS AVAILABLE

Quan.	ITEMS	Code	Unit Price	Quan.	ITEMS	Code	Unit Price
	MEDICAL OUTLINE SERIES				**OTHER BOOKS** *(Cont'd.)*		
	Urology	611	$ 8.00		Neurophysiology Study Guide	600	6.00
	Psychiatry	621	8.00		Hospital Pharmacy Journal Articles	799	10.00
	Cancer Chemotherapy	631	8.00		Psychosomatic Medicine Current J. Art.	788	12.00
	Otolaryngology	661	8.00		Hodgkin's Disease Journal Articles	515	12.00
	OTHER BOOKS				Lithium & Psychiatry Journal Articles	520	15.00
	Practical Nursing Examination - Vol. 1	711	4.50		Human Cytomegalovirus Journal Articles	522	15.00
	English-Spanish Guide for Med. Personnel	721	2.50		Immunosuppressive Therapy Journal Art.	526	20.00
	Multilingual Guide for Medical Personnel	961	2.50		Skin, Heredity & Malignant Neoplasms	744	20.00
	ECG Case Studies	003	7.00		Testicular Tumors	743	20.00
	Endocrinology Case Studies	008	10.00		Neoplasms of the Gastrointestinal Tract	736	20.00
	Infectious Disease Case Studies	011	7.00		Human Anatomical Terminology	982	3.00
	Cutaneous Medicine Case Studies	014	7.00		Critical Care Manual	983	10.00
	Respiratory Care Case Studies	019	7.00		Clinical Diagnostic Pearls	730	4.50
	Emergency Room Journal Articles	795	8.00				
	Outpatient Services Journal Articles	794	8.00				
	Radiology Typist Handbook	981	4.50				
	Surgical Typist Handbook	991	4.50				
	Medical Typist's Guide for Hx & Phys.	976	4.50				
	Handbook of Medical Emergencies	635	7.00				
	Introduction to Blood Banking	975	8.00				
	Cryogenics in Surgery	754	24.00				
	Tissue Adhesives in Surgery	756	24.00				
	Acid Base Homeostasis	601	4.00				
	Radiological Physics Exam. Review	486	10.00				
	Benign & Malignant Bladder Tumors	932	15.00				
	Surgery Self-Assess. Current Knowledge	250	10.00				
	Selected Papers in Inhalation Therapy	523	10.00				
	Practical Points in Gastroenterology	733	7.00				
	Concentrations of Solutions	602	3.00				
	Diagnosis & Treatment of Breast Lesions	748	15.00				
	Hospital Housekeeping Journal Articles	790	8.00				
	Hosp. & Inst. Eng. & Maintenance J. Art.	793	8.00				
	Hosp. Security & Safety Journal Articles	796	8.00				
	Institutional Laundry Journal Articles	789	8.00				
	Hosp. Electronic Data Process. J. Art.	791	8.00				
	Ambulance Service Journal Articles	517	10.00				
	Blood Banking & Immunohemat. J. Art.	798	10.00				
	Introduction to the Clinical History	729	3.00				
	Illustrated Laboratory Techniques	919	10.00				
	Blood Groups	860	2.50				
	Outpatient Hemorrhoidectomy Lig. Tech.	752	12.50				

Prices subject to change